Sediment Transport - Flow and Morphological Processes

Editor

Aleix Campo

scitus
academics

Sediment Transport - Flow and Morphological Processes
Edited by **Aleix Campo**

ISBN: 978-1-68117-535-5
Library of Congress Control Number: 2016942935

SCITUS Academics LLC,
www.scitusacademics.com
Box No. 4766, 616 Corporate Way,
Suite 2, Valley Cottage,
NY 10989

Printed in United States of America on Acid Free Paper

Preface

Sediment refers to the conglomerate of materials, organic and inorganic, that can be carried away by water, wind or ice These particulates are typically small, with clay defined as particles less than 0.00195 mm in diameter, and coarse sand reaching up only to 1.5 mm in diameter. However, during a flood or other high flow event, even large rocks can be classified as sediment as they are carried downstream. Sediment is a naturally occurring element in many bodies of water, though it can be influenced by anthropogenic factors. Sediment transport is applied to solve many environmental, geotechnical, and geological problems. Measuring or quantifying sediment transport or erosion is therefore important for coastal engineering. Movement of sediment is important in providing habitat for fish and other organisms in rivers. Therefore, managers of highly regulated rivers, which are often sediment-starved due to dams, are often advised to stage short floods to refresh the bed material and rebuild bars. Geologists can use inverse solutions of transport relationships to understand flow depth, velocity, and direction, from sedimentary rocks and young deposits of alluvial materials.

Sediment Transport - Flow and Morphological Processes provide information on basic and advanced flow mechanisms including turbulence and movement of particles in water. Examples of computational procedures for sediment transport and morphological changes are presented. This book put together recent developments on sediment transport and morphological processes. Sediment transport is important in the fields of sedimentary geology, geomorphology, civil engineering and environmental engineering. Knowledge of sediment transport is most often used to determine whether erosion or deposition will occur, the magnitude of this erosion or deposition, and the time and distance over which it will occur.

Table of Contents

Chapter 1

A Comparative Study on Fuel Consumption Reduction Effects of Eco-Driving Instructions Strategies

Shuichi Matsumoto[1], Taehwi Park[2], Hironao Kawashima[3]

[1]Faculty of Science and Technology, Keio University, 3-14-1 Hiyoshi, Kohoku-ku, Yokohama 223-8522, Japan

[2]Faculty of Medicine, Asahikawa Medical University, 1-1 Midorigaokahigashinijo, Asahikawa 078-8510, Japan

[3]Co-Mobility Society Research Center, Keio University, 7-1 Shinkawasaki, Saiwai-ku, Kawasaki 212-00322, Japan

ABSTRACT

Eco-driving is expected to contribute to the efforts to fight against global warming by reducing carbon dioxide emissions from vehicles. It is the driving technique that enables low fuel consumption mainly by accelerating gently at the start and avoiding frequent accelerating and decelerating. The effectiveness of eco-driving has been assessed based on a single vehicle; however, it remains to be investigated on multiple vehicles. This paper assesses the effects of eco-driving when following other vehicles on open roads under two different instructions (Japanese eco-driving and German eco-driving) to reduce fuel consumption based on the experiment environment by integrating a driving simulator and a traffic simulator.

KEYWORDS

ITS, Traffic flow, Global environmental problems

1. INTRODUCTION

The Kyoto Protocol, adopted in 1997, required Japan to reduce its greenhouse gas emissions over the period from 2008 to 2012 to 6 % below the level in 1990. In addition, in June 2009, Japan announced a mid-long term target to reduce its greenhouse gas emissions by 25 % from the 1990 levels by 2020. The breakdown of carbon dioxide emissions in Japan indicates that the emissions from transportation account for 20 % of the total, and it is mostly from vehicles. In such situations, eco-driving is attracting attention as one of the ways to reduce carbon dioxide emitted by vehicles.

Eco-driving is considered to be effective in improving the fuel efficiency and is being promoted in various ways. One of the representative activities in Japan is "10 tips for fuel-conserving Eco-Driving" promoted by Team Minus 6 % [1]. "10 tips for fuel-conserving Eco Driving" recommends drivers techniques such as a soft starting with gentle acceleration, called "e-start", and driving without excessive accelerating and decelerating. In this paper, such driving techniques are collectively referred to as Japanese eco-driving. On the other hand, in Europe especially in Germany, the driving technique to accelerate quickly to reach the fuel efficient range as fast as possible, which differs from Japanese eco-driving, is being widely implemented as eco-driving [2]. This driving technique is called German eco-driving. The biggest difference between Japanese and German eco-driving is how to accelerate when the vehicle starts moving. This study aims to evaluate quantitatively the difference in the fuel consumption rates during driving between Japanese eco-driving and German eco-driving.

2. POSITION OF THIS STUDY

The effect of Japanese eco-driving in reducing fuel consumption rates has been verified in various actual driving situations, including the comparison of before and after education for transport drivers such as truck drivers [3, 4] and the comparison of before and after instruction of eco-driving for general drivers [5]. However, there have been few studies of comparison performed in the same conditions.

With regard to German eco-driving, Beusen et al. [6] reported that the fuel consumption was improved after a driver education course on eco-driving was given compared to that of before, through obtaining the driving data of 8 subjects.

As a comparative study on the effects of Japanese eco-driving and German eco-driving, the author et al. [7] analyzed the differences of the effects between Japanese eco-driving and German eco-driving on the fuel consumption rates based on the data of actual vehicle experiments. In this study, the fuel consumption rates were estimated based on the distributions of velocity-acceleration in normal driving, driving with Japanese eco-driving and with German eco-driving. In addition, Kato et al. [8] held an eco-driving test-ride event with actual vehicles to compare the difference between eco-driving while paying attention to gentle acceleration (equivalent to Japanese eco-driving in this paper) and while paying attention to speedier acceleration (equivalent to the German eco-driving in this paper), which proved that both techniques of eco-driving improved the fuel efficiency compared to normal driving.

As a study using a driving simulator (hereafter called "DS"), we can look at the study by Hiraoka et al. [9]. In this study, an experiment was conducted with the condition to drive following the preceding vehicle in JC-08 mode. It concluded that German eco-driving improves fuel efficiency more than Japanese eco-driving does when compared to normal driving. Furthermore, to conduct an empirical study on open roads with traffic lights is cited as a future challenge.

Therefore, this study analyzed while focusing on the vehicle behavior between two intersections with traffic lights, and compared Japanese eco-driving and German eco-driving about the effect to improve fuel efficiency and the driving behaviors over short distances on open roads in DS, which has been rarely discussed in detail until now.

3. EXPERIMENTAL OVERVIEW

3.1 Experiment Environment

In this study, an experiment was performed under an environment with simulated intersections in a virtual space, created by integrating a DS by Keio University and a traffic simulator “AIMSUN” [10] (hereafter called “TS”) (hereafter called “TS/DS integration”). The DS was jointly developed by Keio University and the National Institute for Land and Infrastructure Management [11], and is made by MITSUBISHI PRECISION CO., LTD. The TS is made by TSS-Transport Simulation Systems, S.L.(TSS).

Our study group has performed computer experiments using a TS in the study into the effects of signal control and street parking on smooth traffic flow [12] and in the review and evaluation of public transportation operational procedures [13]. And the study also used a DS in our experiments into the effects of information provision in consideration of human factors of drivers [14, 15]. Experiments with a TS have the advantage that we can evaluate traffic flow through a vehicle group, using a following model of generalized vehicles. On the other hand, experiments with a DS have the advantage that vehicle control data can be quantified based on each driver's driving characteristics. Incorporating these advantages of TS and DS, we created the TS/DS integration to quantify the driving action in a heavy traffic situation or under complicated signal control. This TS/DS integration is for conducting the driving experiment using a DS in the experimental scenario which is developed through use of a TS. In addition, the data output of travel time, delayed time, stopping time and average velocity can be obtained from the TS, and that of acceleration, velocity, engine speed, accelerator pedal input, brake pedal input and the position of the vehicle can be obtained from the DS.

The DS used in this study is composed of eight image screens, front, left and right fronts, left and right sides, left and right rear, and backside (Refer to Fig. 1). A driving image from 10 projectors is projected to these screens. A 360-degree field of vision is provided to the driver in the driver seat. The left and right side mirrors and the rearview mirror are the same type as those for actual vehicles, and they capture indirectly the image on the reverse side of the screen. Furthermore, swing equipment with six axes is attached to the vehicle body. All the equipment is collectively controlled using the main calculator.

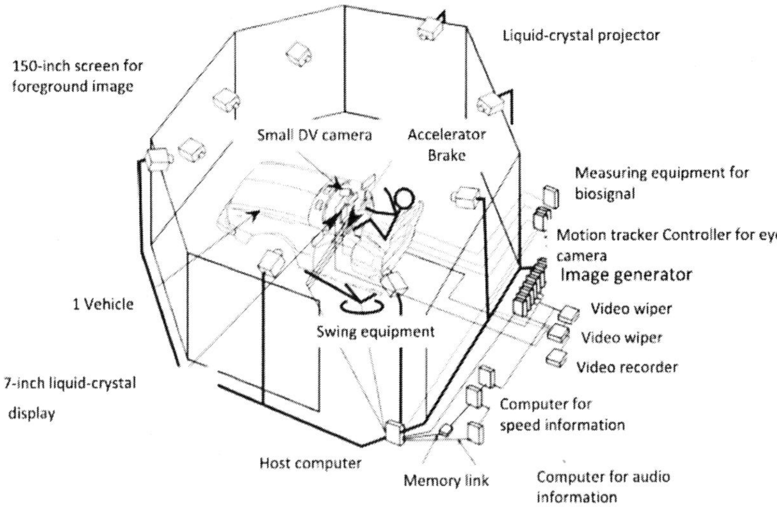

Figure 1. General picture of driving simulator

In this study, an experimental scenario of a straight open road was developed using a TS. The scenario begins with the situation where a vehicle stopped at a traffic light starts moving. It drives straight for 300 m, then must stop at a signalized intersection, and drives straight for 300 m. Each subject got into a DS controlled vehicle which had set that there were vehicles driving in front and at the back. (Refer to Fig. 2). The preceding and following vehicles ran according to the vehicle behavior model programmed in the TS [10], and their speed limit was set at 50 km/h. The main TS parameters, which affect drivers of both preceding and following vehicles, were set as in Table 1. As the vehicular swept path of the preceding vehicle, the relation between velocity and distance is shown in Fig. 3.

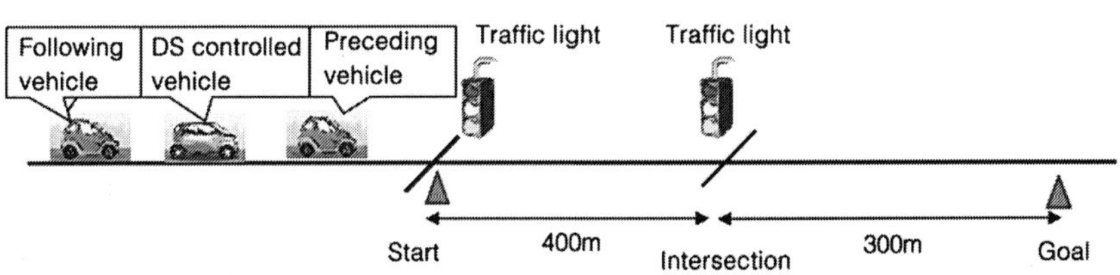

Figure 2. Summary of experiment scenario

Table 1. Main TS parameters

Item (unit)	Mean value (standard deviation)
Maximum desired speed (km/h)	50(0)
Maximum acceleration (m/s^2)	3.0(3.0)
Maximum speed (m/s^2)	6.0(0.5)
Minimum inter-vehicular distance (m)	1.0(0.3)

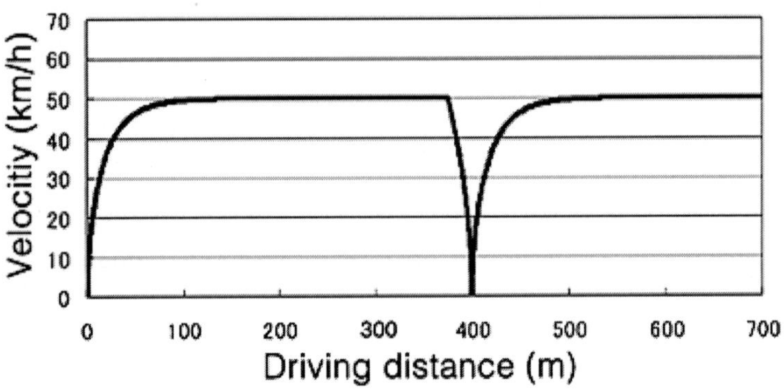

Figure 3. Relationship between velocity and driving distance of preceding vehicle

3.2 Calculation of Fuel Consumption

In this study, the fuel consumption amount calculated based on the map shown in Fig. 4, which is in the same manner as the study by Hiraoka et al. [9]. This map is included in the vehicle motion simulation software, CarSim. It shows the instantaneous fuel consumption amount (kg/s) for engine speed and an operation amount on the accelerator of the ordinary motor vehicle with a 2.5-l gasoline engine. We calculate the fuel consumption rate km/l from this obtained amount, the standard density of gasoline 0.76 kg/l and the vehicle velocity km/h.

Figure 4 shows that in the area with very low accelerator pedal input, the fuel consumption amount does not change remarkably even if the engine speed is increasing. When a vehicle is in a coasting state or a braking state, it is usually included in this area. In the area with high accelerator pedal input, the fuel consumption amount increases as the engine speed increases However, it is rare to accelerate with full force, with the exception in some particular circumstances, for example, when merging onto an expressway or driving up a steep slope. The area with the accelerator pedal input about 0.2 to 0.6, called partial throttle area, can be the area where both of the accelerator pedal input and the engine speed become very influential. In particular, an engine speed exceeding 3,600 rpm will tend to sharply increase the fuel consumption amount. In this study, the instantaneous fuel consumption amount (kg/s) is calculated using the accelerator pedal input and the engine speed obtained with a DS.

Though various models are advocated as fuel consumption amount estimation models, the model in this study was chosen because it could provide the driving behavior of drivers and vehicle control data of the DS based on the behavior, enabling the calculation of the fuel consumption amount using the accelerator pedal input and the engine speed (Fig. 4).

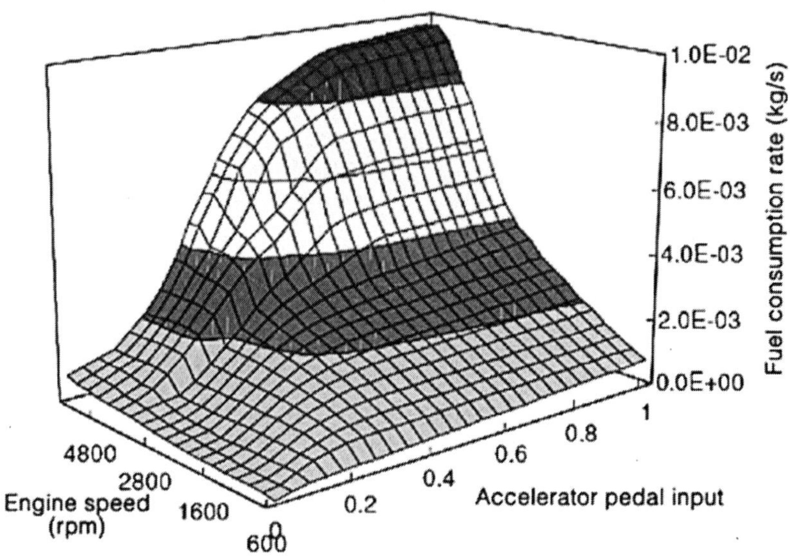

Figure 4. Fuel consumption rate map

3.3 Summary of Attribution of Subjects and Instructions

The subjects are five males (age: 22 to 24 years, average age: 22.6 years, standard deviation: 0.8 years). They filled out the questionnaire about their attribution such as age and experience in eco-driving before the experiment. Table 2 summarizes the attributes of each subject who participated in this experiment.

Table 2. Subject's attribution

ID	Age	Driving experience	Driving frequency
A	23	4 years 10 months	1–2 times per week
B	22	2 years 2 months	Once in every 2 months
C	24	2 years 11 months	1–2 times per month
D	22	2 years 11 months	1–2 times per week
E	22	4 years 1 month	1–2 times per month

The subjects drove a total of 15 times, 5 times each with normal driving, Japanese eco-driving, German eco-driving in this order. The reason for this order is because it was considered that Japanese and German eco-driving would affect normal driving if the test with normal driving was carried out after those with Japanese and/or German eco-driving. As for the order of Japanese and German eco-driving, it is preferable that the order be randomly set for each subject, in consideration of the order effect. In the study, the order was set to avoid the psychological bias in subjects on safety issues such as getting too close to the preceding vehicle if German eco-driving precedes Japanese one. As only 5 subjects participated in the study, the number of driving times was increased so as to obtain more data samples for analysis. As a future challenge, further validation of the experiment with more subjects is necessary. The subjects received instruction on the driving technique and practiced how to drive, before the first test of each driving technique. The instruction is summarized as below.

- Normal driving

Drive in the same way as they do in their daily lives.

- Japanese eco-driving

1 Press on the accelerator gently using creep phenomenon to start moving (approximately 20 km/h in 5 s).

2 Minimize accelerating and decelerating as much as possible and maintain steady acceleration.

3 Release the accelerator early and actively utilize engine braking.

- German eco-driving

1 Accelerate as fast as possible (but not too quickly) to reach the fuel efficient velocity range (40 to 80 km/h).

2 Release the accelerator after reaching the fuel efficient velocity range and utilize coasting actively.

4. EXPERIMENTAL RESULTS

4.1 Comparison of Travel Time

In this experiment, the travel time from the starting point to the goal in experimental scenario of Fig. 3 was compared between the normal driving, Japanese eco-driving and German eco-driving. The result is shown in Fig. 5.

When the result went through one-way analysis of variance, the travel time with Japanese eco-driving is longest of all and is longer than that with normal driving by 1.8 % (1.4 s), and with German eco-driving by 2.1 % (1.8 s) ($p < 0.1$), as shown in Fig. 5. Why the travel time with Japanese eco-driving is the longest of all is considered to be due to "e-start", one of the characteristics of Japanese eco-driving, which has gentle acceleration. On the other hand, there is no statistically significant difference in the travel time between German eco-driving which accelerates as fast as possible and normal driving. Based on these findings, the comparisons of German eco-driving and Japanese eco-driving with regard to the way of accelerating reveal that "e-start" used in Japanese eco-driving affects travel time more.

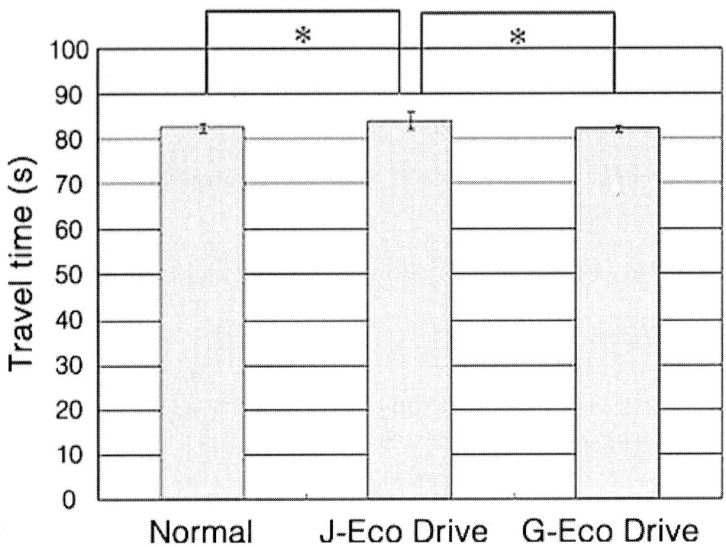

Figure 5. Comparison of travel time (n = 25)

4.2 Comparison of Fuel Consumption Rate

In this chapter, the fuel consumption rates are compared among each of the driving techniques. They are calculated using the fuel consumption rate map, shown in Fig. 4, based on the accelerator pedal input and the engine speed obtained from the DS. Firstly, the results of the fuel consumption rates for each driving technique are shown in Fig. 6.

When the result went through one-way analysis of variance, the fuel consumption rate with German eco-driving was 4.1 % (0.48 km/l) higher than that with normal driving, and that of Japanese eco-driving was 0.87 % (0.1 km/l) lower than that of normal driving, as shown in Fig. 6. However, no statistically significant differences were observed in these results.

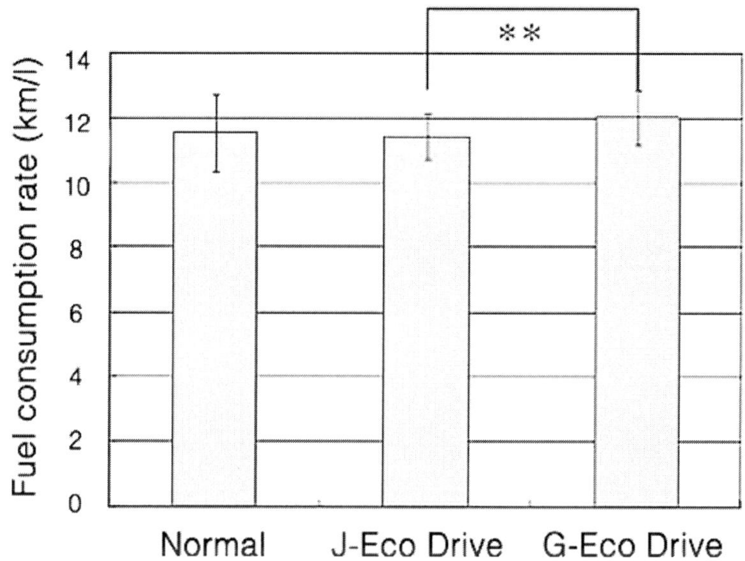

Figure 6. Comparison of fuel consumption rate of each driving technique ($n = 25$)

It is shown that the fuel consumption rate with German eco-driving is 5.1 % (0.56 km/l) higher than that of Japanese eco-driving ($p < 0.05$).

In general, it is said that driving with Japanese eco-driving would reduce the fuel consumption rate compared to normal driving. However, the results of this experiment do not indicate that Japanese eco-driving, which minimizes accelerating and decelerating, will reduce the fuel consumption rate when the intervals between intersections are short. Next, we define the state with 1) no accelerator pedal input nor brake pedal input and 2) a negative acceleration as the coasting state. The ratio of coasting state, which is the proportion of the driving time in the coasting state to the total travel time, is calculated. The result is shown in Fig. 7. Figure 7 shows that the ratio of coasting state with German eco-driving is by far the highest, and that with Japanese eco-driving is the lowest ($p < 0.01$).

Additionally, as an example of driving distance and accelerator pedal input for each of normal driving, Japanese eco-driving and German eco-driving in the scenario for this experiment, the driving data of the third test by Subject A is shown in Figs. 8, 9and 10.

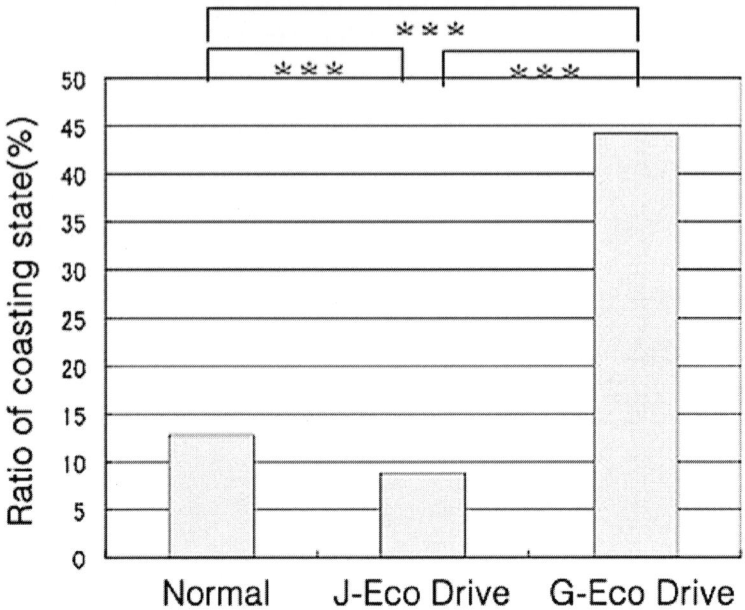

Figure 7. Comparison of ratio of coasting state of each driving technique ($n = 25$)

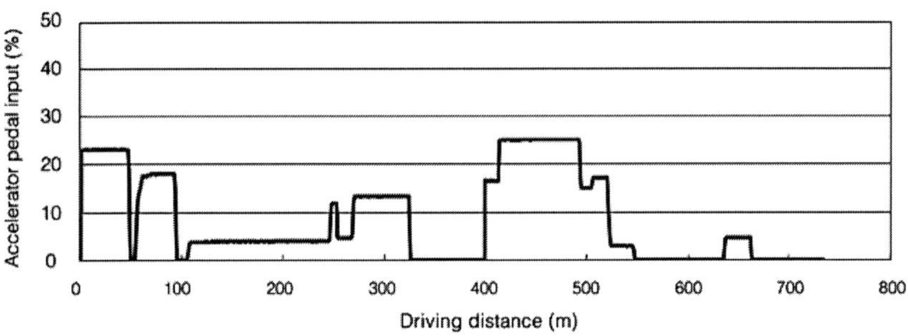

Figure 8. Example of accelerator pedal input and driving distance (Normal driving)

As shown in Fig. 9, with Japanese eco-driving, the duration the accelerator is pressed is long even though the accelerator pedal input is small because Japanese eco-driving utilizes “e-start”, which made its ratio in the coasting state small. This can be the reason why the fuel consumption rate with Japanese eco-driving was worse than that with German eco-driving.

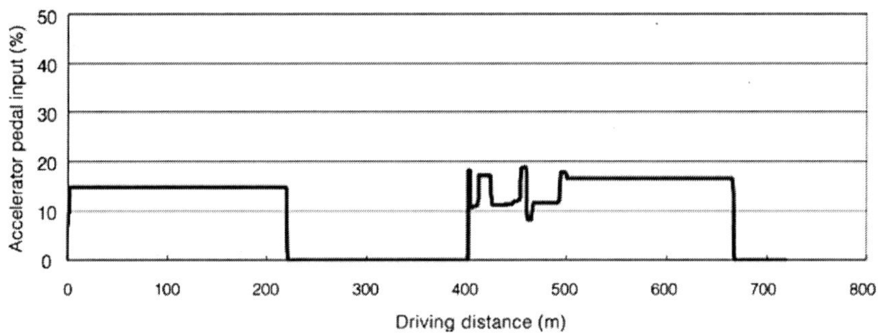

Figure 9. Example of accelerator pedal input and driving distance (Japanese eco-driving)

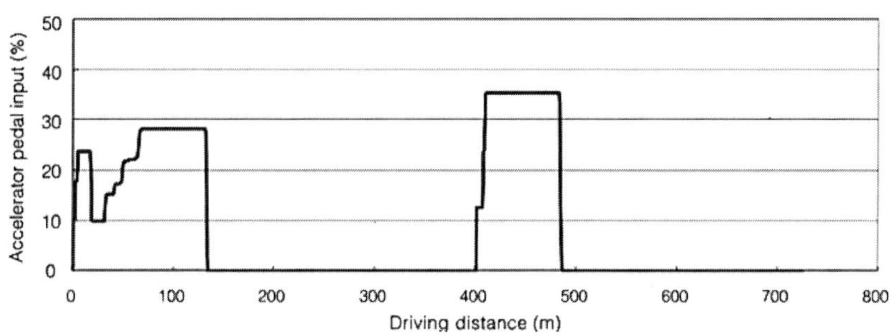

Figure 10. Example of accelerator pedal input and driving distance (German eco-driving)

On the other hand, as shown in Fig. 10, the fuel consumption rate with German eco-driving right after the start is higher because the accelerator pedal input is high at the start. It is considered that its fuel consumption rate is better than those of the other driving techniques because when it goes into the coasting state, it requires little fuel consumption after that.

Thus, this study suggested that Japanese eco-driving would not always have an effect to reduce the fuel consumption rate compared to normal driving were obtained under the condition that the vehicles should stop two times at traffic signals during their 700-m drive.

In the future, we intend to study quantitatively the effect of eco-driving on the whole traffic flow in the case of high volume traffic. In addition, we intend to study in detail the difference between those subjects who drive with eco-driving techniques routinely and those who don't, by performing an experiment with an increased number of subjects. Moreover, we plan to analyze the changes in fuel consumption amount depending on traffic situations, distance between traffic signals, instructions on eco-driving and so on. Furthermore, we plan to try a new condition of experiment where a DS controlled vehicle is driven while reading the movement of the preceding vehicle in order to drive in the most fuel efficient manner.

REFERENCES

1. The Eco-Drive Promotion Liaison Committee: 10 tips for fuel-conserving Eco Driving. (2005)
2. Ford-Werke GmbH: Ford Eco-Driving Kompakt, http://www.ford-eco-driving.de/
3. Maji, H., Kasuga, N., Ishi, T., Daisho, Y.: Effects of eco-drive activity on improvement in fuel economy and reduction in traffic accidents. Proc. JSAE **38**(3), 99–104 (2007)
4. Saito, A., Takada, H.: The relationship between eco-driving patterns of light duty freight vehicles and the improvement of fuel economy in real traffic conditions. Proceedings of JSAE before lecture presentation, No.99–08. pp. 27–30 (2008)
5. Miyasaka, T., Taniguchi, M., Sanbuichi, H.:Potentiality of effects by driving operation for fuel conservation to automatic transmission car. Proceedings of JSAE before lecture presentation, No.80–05. pp.13–16 (2005)
6. Beusen, B., Denys, T.: Experiences using an onboard logging device with CAN-interface, GPS, GPRS, and Wi-Fi for surveying long term driving behavior. Proceedings of 15th World Congress on Intelligent Transport Systems, CD–ROM. (2008)

7. Matsumoto, S., Kobayashi, I., Nakamura, A., Kawashima, H.: Economic evaluation of influence on traffic flow by eco-driving. Annual report of JSTE, Japan Society of Transportation Economics. pp. 169–178 (2007)

8. Kato, H., Kobayashi, S.: Factors contributing to improved fuel economy in eco-drive. J. JSAE **62**(11), 79–84 (2008)

9. Hiraoka, T., Terakado, Y., Matsumoto, S., Yamabe, S.:Quantitative evaluation of eco-driving on fuel consumption based on driving simulator experiments. Proceedings of 7th World Congress on Intelligent Transport Systems. pp. 163–168 (2008)

10. Babazono, K.: Introduction of microsimulation environment tool, AIMSUN NG. IPSJ SIG Technical Reports ITS **2005**(61), 13–20 (2005)

11. Munehiro, Y., Yamazaki I., Daimon T., Arizumi M.: A counter-measure for crossing collisions based on analysis of human error using driving simulator. Infrastruct. Plan. Manag. JSCE/Proc. **34** (2006), CD–ROM

12. Nakamura, A., Matsumoto, S., Todoroki, T.: Measuring external costs of on-street parking near the bus-stop: the case of Chiba. Annual report of JSTE, Japan Society of Transportation Economics, pp.159–168 (2008)

13. Matsumoto, S., Kumagai, Y., Kawashima, H.: A study on the traffic signal strategy in city traffic in consideration of streetcar. Infrastruct. Plan. Manag. JSCE/Proc **21**(3), 677–685 (2004)

14. Matsumoto, S., Tozawa, T., Hiraoka, T., Yamabe, S., Kawashima, H.: A fundamental study on the effect of eco-driving on following vehicles. Infrastruct. Plan. Manag. Proc. D3 **67**(5), I_1023–I_1028 (2012)

15. Hiraoka, T., Terakado, Y., Matsumoto, S., Yamabe, S.: Effect of eco-driving instruction and fuel consumption meter on fuel consumption reduction. Hum. Interf. Soc. J. **12**(1), 71–78 (2011)

reduce parking search traffic particularly at times of unexpectedly high parking pressure. The real-time occupancy model presented in this article does not aim to indicate the occupancy of single parking lots, but to provide a sufficiently accurate description and reliable prediction of the occupancy at the destination. Three existing real-time indicators were tested for their predictive power: (i) location data of mobile-phone parking; (ii) counts of car parks in the vicinity; and (iii) traffic flow volumes. Experimental results indicated that models using mobile-phone parking data have the potential to improve predictions of the occupancy rate by detecting exceptional deviations from the average day curve.

KEYWORDS

ITS, Telematics, Parking management, Short term parking, Mobile phone parking, Real-time occupancy model

1 INTRODUCTION

Driver assistance systems (DAS) became very popular in recent years. The automobile industry and academic researchers spend much effort to develop such systems [16]. They can detect certain driving situations and provide car drivers with real-time information on different aspects such as travel time [1] and parking facilities. The parking process can involve several annoying tasks: finding a free parking space, manoeuvring into a narrow space, finding a ticket machine or filling in a prepaid ticket etc. As a result, there is growing interest in parking assistance systems.

There are several techniques, which provide drivers with real-time information on the occupancy of car parks and guide them to the next free parking space. Counting systems like traditional induction loops and video detection [7] as well as secured wireless networks are available in car parks to record entering and existing vehicles. The degree of occupancy (or number of free spaces) is displayed on variable message signs (VMS) and drivers are guided to available parking facilities with directional signs [15]. A recent contribution is “smart parking”, which has been implemented in many European and Japanese cities to improve the speed and efficiency of locating, reserving and paying for parking [18]. But even inside a car park assistance may be required to find a free space. There are vision-based systems [13] and wireless sensor networks (WSN), which have been developed and tested, e.g., by Seong-eun et al. [20], Jaeseok et al. [10], Tang et al. [23] and Gongjun et al. [8]. They inform drivers about free parking spaces and route them directly to a free place.

Many drivers would welcome analogous information services for on-street parking spaces. This task is more difficult, because entries, exits, and parking spaces are less well defined and inappropriate for counting. On-street parking vehicles can be detected with panoramic street images taken with different cameras. Hirahara and Ikeuchi [9] used a laser-ranger finder and a line-scan camera taking epipolar-plane images (EPIs) for this task. This method delivers only static snap-shots. Most attempts to deliver real-time information about parked vehicles on-street rely on roadside sensors. For example, Berenger Vianna et al. [3] as well as Basu and Little [2] developed a system with linked parking meters to find free spaces. Parking meters were equipped with infra-red sensors to identify free and occupied parking spaces; the parking meters were connected with each other via radio frequency transceivers. Free parking meters send their position to the network and passing cars can receive this information as well. BMW developed a similar system in the project 'stadtinfoköln' with about 140 connected parking meters in Cologne [14]. This information service is available for 19 German cities and can be used for free via internet.

Wired sensors networks are impractical to cover large areas that lack a suitable electrical infrastructure because of high deployment and maintenance costs [21]. An alternative option for detecting the occupancy of on-street parking spaces are wireless sensor networks with autonomous sensor nodes [7]. Most of these systems use ultrasonic sensors or induction loops. The devices communicate via radio frequency transmitters; the occupancy information can be accessed through mobile devices [12] or being displayed on VMS. Cenerario et al. [4] took a different path. They employed protocols that vehicles can use to share information by means of V2V communication; it could for instance be used to share information about available parking spaces. A similar community-based system is currently tested in Vienna [22]. It enables users to indicate (release) free on-street parking spaces and to find these spaces via a mobile app. This system would however require a high share of market penetration to provide a complete picture, which can only be achieved in the long run.

An accurate detection of on-street parking spaces relies invariably on area-wide infrastructure on street or within vehicles, which causes high costs for implementation and maintenance. Such systems are also controversially discussed, because they aim at navigating drivers as quickly as possible to the next free parking space, that way making car use more convenient. It is commonly accepted that car use should not be further encouraged, in particular not in the inner districts of large cities. This argues for an information system, which informs road users early about the current and forthcoming parking situation at the destination—preferably prior to departure, when there are still more options available to adapt the behaviour.

Many drivers are able to roughly estimate the availability of short-term parking spaces at the destination based on experience. However, situational influences such as holidays, building sites, and special events introduce variability to the parking pressure, what encourages people to try their luck. As a result, parking search traffic amounts to 25 % of total traffic in the inner districts of Vienna [6]. Rye et al. [19] emphasises that managing information about parking availability should be part of demand management. A more reliable forecast of the occupancy could support the use of alternative options (public transport, Park & Ride, garage at the destination) and reduce parking search traffic at the destination. The self-regulating effect could be reinforced by variable parking fees depending on the demand. Such schemes are already in operation, e.g. in Amsterdam [17].

The city of Vienna offers an SMS payment service for short-term parking fees, which is called “HANDY Parken”. More than 40 % of the short-term parking tickets are paid by this means. The success of HANDY Parken in Vienna opens the way for a solution, which is independent from on-street sensors. Each electronic ticket generates information about the duration of the parking process and the approximate location of the vehicle via cell location. The approach is based on the assumption that these data provide a probabilistic indicator for the number of parked vehicles, thereby enabling an estimation of the total occupancy. A real-time readout procedure of position data of HANDY Parken does not exist yet, but has to be developed.

2 OBJECTIVES

The objective of the research project this article reports on was to develop a real-time model for the occupancy of short-term parking zones, which works without roadside infrastructure, but only uses existing real-time data sources (Fig. 1). Three different kinds of data were tested for their predictive power:

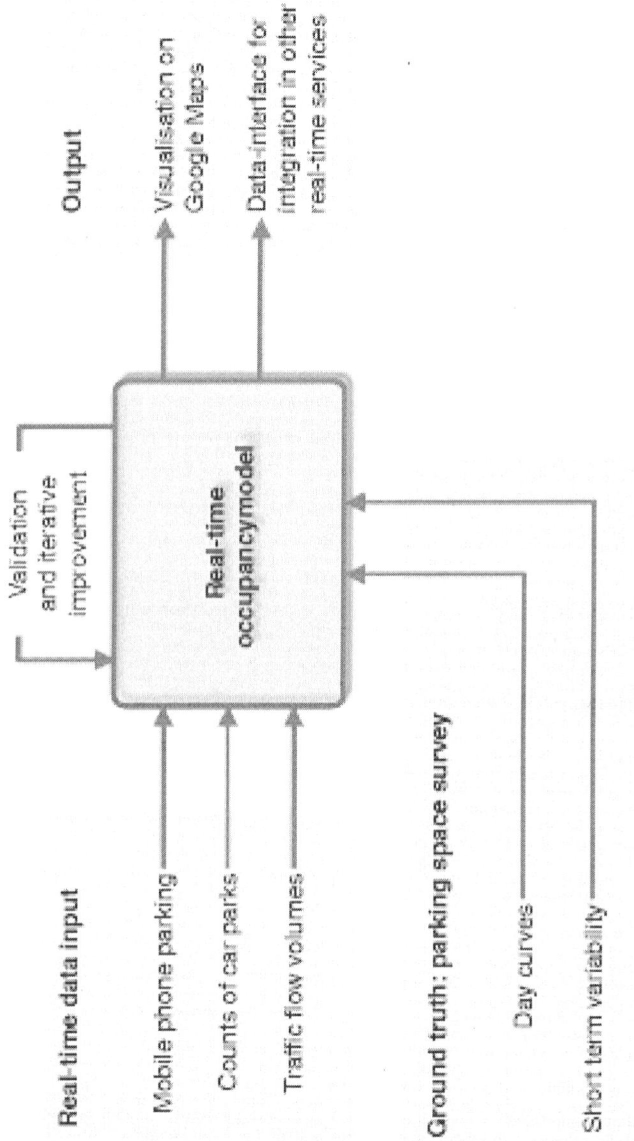

Figure 1. Data flow in the real-time occupancy model

- position data of short term parking customers, who buy an electronic parking ticket on their mobile phone via the HANDY Parken service;

- counts of short-term parking customers in car parks in or close to the target area;

- traffic flow volumes derived from around 300 automatic counting sites in Vienna.

We focussed on these data sources, because they are already available or could be gained at low cost. Moreover, this article is only about (i) testing for the predictive power of these data sources, (ii) developing a suitable model specification, and (iii) evaluating the performance of the model. The development of a physical model is not part of this article. The resulting model shall not exactly map the occupancy of single parking lots, but provide a sufficiently accurate description and reliable prediction of the parking pressure in the target area.

Vienna provides optimal conditions for this attempt due to the high share of mobile phone parkers. Parking meters are not available in Vienna and sending an SMS is apparently considered much more convenient than filling in a paper ticket. As a result, more than 40 % of parking tickets is paid by this service, what corresponds to more than a million SMS tickets every month. This makes HANDY Parken the most commonly used mobile service in Austria.

The mobile phone data are generated only during the validity period of the short-term parking zone and arrive just in time, which limits their usability for a forecast. Therefore, two further real-time data sources were tested in this study, which are available all time and may increase the reliability of forecasts. Both data sources are state of the art, but their use as real-time indicator of the occupancy of short-term parking zones remains to be proven.

The real-time occupancy model was developed and calibrated with data of an empirical parking space survey in two test areas. The most challenging tasks were to find out, (i) if the extraction of position data of HANDY Parken customers is technically and legally feasible and (ii) if the three aforementioned indicators can predict the real-time occupancy of the short term parking zones. The expected result is an online prototype, equipped with a visualisation on Google Maps and a data-interface for integration into existing travel information services. Preliminary results were presented at the ITS World Congress 2012 in Vienna.

3 REAL-TIME DATA SOURCES

The mobile phone parking data constitute the most promising indicator of the occupancy model. The HANDY Parken service has an agreement with several mobile network operators (MNOs), the largest of which participated in the research project. Reading out all SMS-tickets booked in a given time interval together with the permitted parking duration is technically and legally feasible. But, we also need to determine the position of the mobile phone from which the SMS was sent. It serves as proxy for the position of the car and is allocated to a particular street section. We aimed to perform this task by determining the coordinates of the active network cell. This procedure is less precise than GPS location, but can be universally applied, not only on smart phones. Due to the high density of network cells in the inner districts of Vienna, we expect a spatial accuracy of 100–200 m. Given that drives roam areas of a similar size when looking for a parking space, a more precise location is presumably not necessary for an information service about the occupancy of on-street parking spaces.

The Austrian Data Protection law under the rule of the European Data Protection law (EU-Dir. 95/46/EC and EU Dir. 2002/58/EC) [5] as well as the Austrian Telecommunication Act (§99 TKG2003) raise several cautious questions regarding privacy concerns: What kind of data retrieval is permitted for the MNO? At which stage of data processing will the aggregation take place? What level of aggregation is sufficient? These questions could not be answered in advance, but had to be discussed during the project. The data protection officers of the MNO finally decided against a data transfer despite an initial agreement. As a result, the HANDY Parken data could be made available for calibrating the model, but will not be available for operation.

The counts of short-term parkers in the car parks were easier to get, because all car parks in Vienna dispose of automatic counting systems. The counts are periodically retrieved and fed into the Viennese parking guidance and information system. However, for the development of an occupancy model it was necessary to distinguish between customers with long term contracts and short term customers. Since the parking information system requires only the total number of free parking spaces for its operation, this distinction is not common state of practice. We achieved an agreement with five of the six car park operators in the test areas. All of them were able to distinguish between long term and short term parkers and provided the counts during the parking space survey.

Traffic flow data could be collected without any difficulty. They are already collected at a central point, where they feed into a transport model, which is in service atwww.AnachB.at. A data exchange protocol has been established, so that we could retrieve the data at any time.

4 PARKING SPACE SURVEY

The parking space survey conducted in the research project served two purposes: (i) to deliver the 'ground truth' for the calibration and validation of the occupancy model; and (ii) to deliver the day curves of those vehicles, which have a permanent parking permission, any other special permission, or no permission at all (fee dodgers). These parkers leave no electronic trace for real-time indication. The survey was conducted in two test areas as shown in Fig. 3 with a total of 3,000 parking spaces. The areas vary in terms of residents-visitors ratio, supply of on- and off-street parking, spatial functions, and demographic structure. The survey was carried out in three periods in February, March and April 2012. It should be noted that the parking fee had been increased between period 1 and 2 from previously 1.20 to 2.00 €/h (the survey also served to estimate the effect of this intervention), so that only the 2nd and 3rd period are suitable for cross-validation under ceteris paribus conditions.

The survey started with a record of all parking supplies. For each available parking space it was indicated, whether it is a regular short term parking space, a special parking space for loading, disabled drivers, or any other specific group. The demand of parking spaces was afterwards observed periodically every 30 min from 8 am to 10 pm. Each parked vehicle was registered along with the type of parking permission in terms of permanent permission, fill-in ticket, or no visible permission. The latter could be fee dodgers or users of the mobile parking service. In order to distinguish between the two, we recorded the license plates of these vehicles and checked them against the HANDY Parken database, which allowed us to determine, which of these cars had a valid electronic ticket during the observation period.

5 OBSERVED OCCUPANCY RATES

The cars parking on short-term parking spaces can be grouped into the following categories according to the type of parking permission:

- permanent permission (mainly residents, but also businesses and other specific groups);
- short term permission in terms of a prepaid fill-in ticket;
- short term permission in terms of an electronic ticket; and
- no permission; these are fee dodgers who forgot to buy a ticket or deliberately try their luck that they are not enforced and punished; the share is 12 % in the total sample; it matches well with the experience of the parking surveillance authority.

The repeated observations of parking cars (every 30 min from 8 am to 10 pm on 11 days) result in 28 × 11 = 308 data points in time for each street section. The objective of the occupancy model is to predict the occupancy rate of an area rather than the number of parked cars. We thus translated the car counts into fractions of available parking spaces occupied by different categories as shown in Fig. 3. The result reveals strong differences between the areas (Fig. 2). The share of short-term parkers is approx. 40 % in the city centre (area 1), whereas it is only 20 % in the residential area (area 2). The day curves are also different. The occupancy in the city centre is highest during the day and again during the opening hours of pubs in the evening, whereas the residential area shows the highest occupancy in the morning before the residents leave and in the evening after they are back.

6 ANALYSIS OF RELATIONSHIPS BETWEEN OCCUPANCY RATES AND REAL-TIME DATA SOURCES

This section includes the tests of linear relationships between the three kinds of real-time data and the occupancy of short term parking zones. The tests were performed using correlation coefficients. The results and conclusions presented in the following hold only true for Vienna. We do not have data outside Vienna, so we cannot discuss the applicability for other cities.

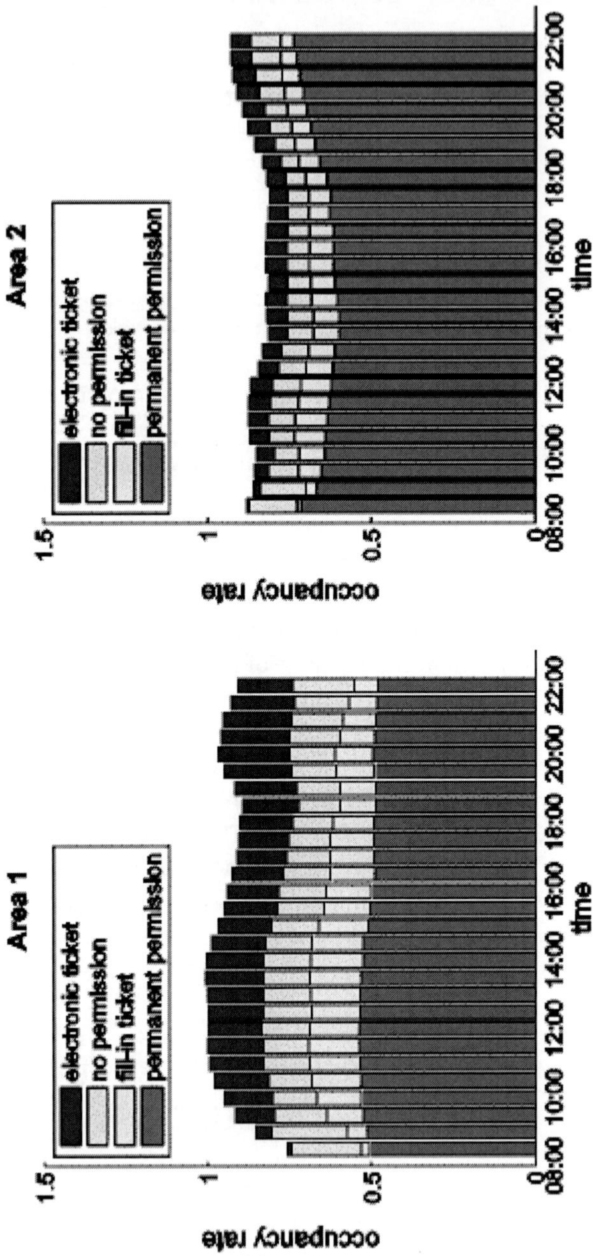

Figure 2. Share between different types of parking permissions as fractions of the estimated number of available parking spaces in 1st district (area 1) and 6th district (area 2)

6.1 Counts of Short Term Parking Customers in Car Parks

Car park counts were collected from six car parks in or close to area 1 and three car parks in or close to area 2 (green triangles in Fig. 3). We computed several correlations of car park data with parking space occupancy. As an illustration, Fig. 3shows the correlations of the inflow of short term parking customers into the car parks with the average parking occupancy of the corresponding street sections. We found no systematic relationship between car park inflow and parking space occupancy, so that the former are no suitable indicator of the latter.

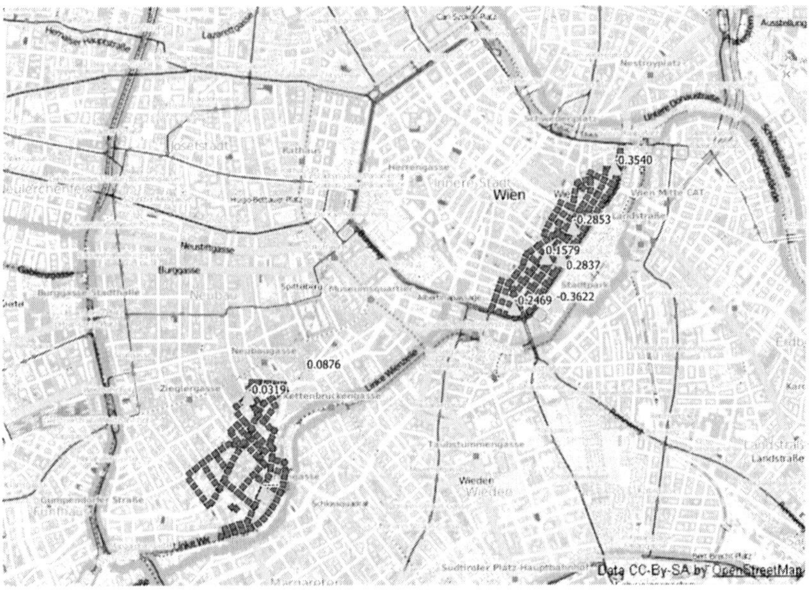

Figure 3. Correlation coefficients of garage inflow with mean parking occupancy in the corresponding area (*right top*: 1st district; *left bottom*: 6th district); red points = street sections, green triangles = car parks, numbers = correlation between inflow of short term parkers into car parks and on-street parking occupancy

Most of the car park operators, who participated in the advisory board of the research project and provided the counts of car park customers, held the view that the garage inflow is just the overflow of on-street parking demand. This view is not confirmed at all by the survey data.

6.2 Traffic Flow Volumes

We examined traffic flow data of 830 major roads in the Vienna region. They were derived from a traffic model operated by the provider of real time traffic information in the Vienna reagion [24]. The Model uses the data of some 300 automatic counting sites as input ([11]). In order to test for the predictive power of the traffic flows, we computed correlation coefficients with mean occupancy rates in both target areas. The results are given in Fig. 4. Although we found some large correlations among the numerous candidates, they have different signs and don't reveal any recognisable geographical pattern. We thus considered the relationship between traffic flow data and parking space occupancy too fragile to be used as predictor in the real-time occupancy model.

Figure 4. Correlation coefficients of traffic flow data with mean parking occupancy rates in area 1 (1st district; *left graph*) and in area 2 (6th district; *right graph*)

6.3 Mobile Phone Parking Data

These data were the most promising candidates for indicating the occupancy of short term parking zones in real time. Our hypothesis was that (i) mobile phone parkers are a congruent subset of all short term parkers, and that (ii) the latter can be added to the number of permanent parkers, which evolves according to its own day curve independent from short term parkers. The mobile network operator involved in the research project refused to provide the location data of mobile phone parkers for privacy concerns (see Section 3). For the vehicles observed in the parking space survey, however, we could retrieve this information from the HANDY Parken database.

Table 1 shows the correlation coefficients of the number of electronic tickets with other kinds of permissions and with the total occupancy rate; it sheds some light on the relationships between these quantities. The strong positive correlation between electronic tickets and fill-in tickets indicates that mobile phone parkers are indeed a congruent subset of all short term parkers (hypothesis 1). However, the assumption of independence between short term parkers and permanent parkers (hypothesis 2) was rejected. The negative correlation with permanent permissions and fee dodgers (without permission) indicates a displacement effect, which may occur in cases when the total demand exceeds parking space supply.

Table 1. Correlation coefficients (*r*) and significance (*p*) between electronic parking tickets and other parking permissions in 1st district (area 1) and 6th district (area 2)

Kind of permission	Area 1		Area 2	
	r	*p*	*r*	*p*
Fill-in tickets	0.515	0.000	0.503	0.000
Permanent permissions	−0.259	0.000	−0.269	0.000
Cars without permission	−0.372	0.000	−0.418	0.000
All other permissions	−0.067	0.241	−0.238	0.000
Total occupancy rate	0.659	0.000	0.115	0.043

This pattern appears similar in both areas. A difference between the areas however appears when it comes to the correlation between electronic tickets and total occupancy rate. Area 1 shows a high positive correlation, whereas area 2 shows a low correlation. The reason is the lower share of permanent parkers in area 1. It causes that the relationship between electronic tickets and total occupancy rate is less adversely affected than in area 2.

Figure 5 shows a linear regression $\widehat{o} = kx + d$ of the total occupancy rate $\widehat{o}$ (last row in Table 1) on the percentage *x* of parking space occupied by cars with electronic tickets. The model explains $r^2 = 43.5\ \%$ of the occupancy rate variance in area 1 and only $r^2 = 1.3\ \%$ of the variance in area 2.

7 DEVELOPMENT OF THE REAL-TIME MODEL

In order to measure the improvement in prediction gained from the real-time information we defined two reference models, which differ in the underlying assumption:

- The constant average model is based on the naive assumption that the occupancy rate stays at a constant average level at any time; it neglects the variability throughout the day. The expected average occupancy rate is simply estimated by

$$\overline{o} = \frac{1}{DT} \sum_d \sum_t o_{t,d} \tag{1}$$

over all days *d* and time intervals *t* from a sample of *D* days each divided into *T* time intervals. The Root Mean Square Error (RMSE) of this predictor is given by the sample standard deviation

$$s_o = \sqrt{\frac{1}{DT} \sum_d \sum_t \left(o_{t,d} - \overline{o}\right)^2} \tag{2}$$

- The average day curve model is based on the slightly more advanced assumption that the occupancy rate follows the same average day curve on all days. The expected occupancy rate in time interval *t* on any day is estimated by

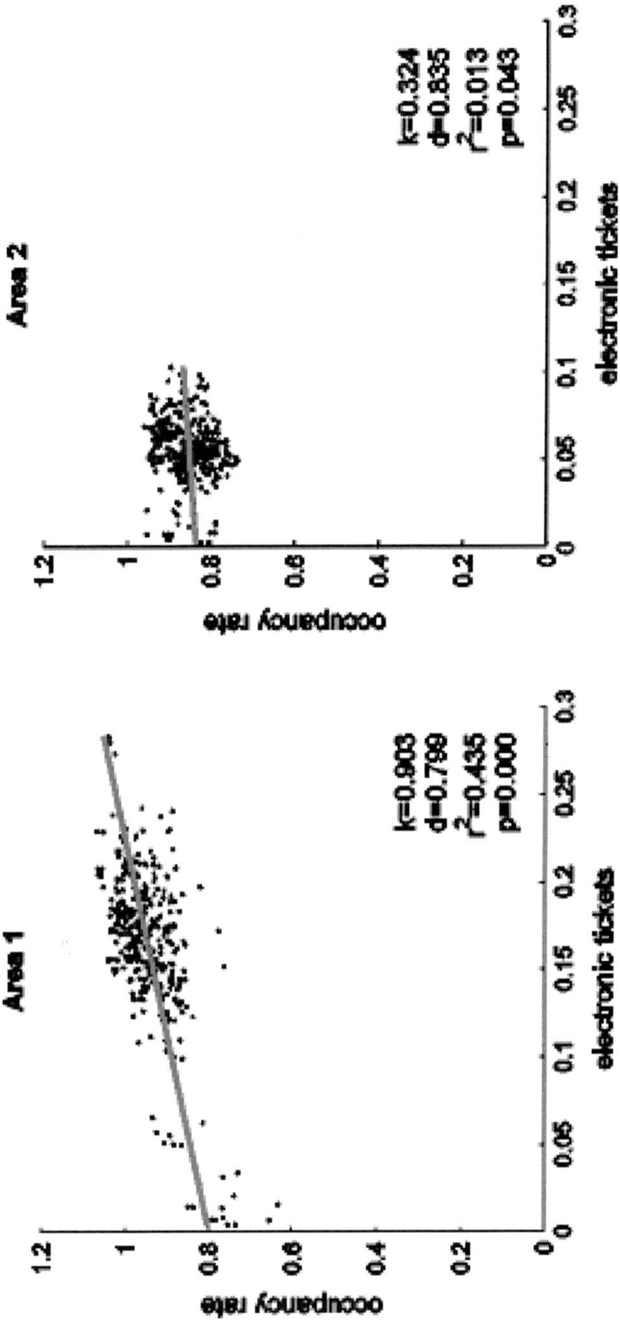

Figure 5. Linear regression of the total occupancy rate on the percentage of parking space occupied by mobile phone parkers in 1st district (area 1) and 6th district (area 2)

$$\overline{o}_t = \frac{1}{D}\sum_d o_{t,d} \tag{3}$$

where $o_{t,d}$ is the occupancy rate on day d in time interval t. This model explains 65.8 % of the variance of the occupancy rate in area 1 and 54 % in area 2, respectively. The resulting day curves are shown in Fig. 2.

We tested the increase of predictive power gained from the real-time information against these models. The car park counts and traffic flow data were excluded from the outset, because no systematic relationship could be identified between these quantities and the parking space occupancy rate (see chapter 6); only the number of purchased electronic tickets was considered in the model. Predictions based on this indicator explain 43.5 % of occupancy rate variance in area 1 and 1.3 % in area 2 (Fig. 5). This is still inferior to predictions based on the average day curve model, which account for 65.8 % and 54 %, respectively (eq. 3). Nonetheless, it is possible that the real-time indicator improves the predictive power, if added to the average day curve model. We tested this possible improvement in three respects:

- improvement of approximation of the occupancy rate in the current time interval;
- improvement of prediction of the occupancy rate of future time intervals;
- improvement of prediction of the occupancy rate in exceptional situations, in which the occupancy rate differs considerably from the usual day curve.

7.1 Approximation of the Occupancy Rate in the Current Time Interval

In order to determine the additional variance explained by the real-time indicator, we calculated the residuals of the average day curve model $\tilde{o}_{t,d} = o_{t,d} - \overline{o}_t$ and, analogously, the residuals of the number of purchased electronic tickets $\tilde{p}_{t,d} = p_{t,d} - \overline{p}_t$. The additional percentage of occupancy rate variance explained by a linear regression model

$$\widehat{o}_{t,d} = k\widetilde{p}_{t,d} + d \tag{4}$$

is $r^2 = 29.3\,\% \times (1 - 65.8\,\%) = 10\,\%$ in area 1 and $r^2 = 3.6\,\% \times (1 - 54\,\%) = 1.7\,\%$ in area 2, respectively. Figure 6 shows a plot of $\widetilde{p}_{t,d}$ against $\widetilde{o}_{t,d}$ for both areas along with the linear regression model according to eq. (4).

7.2 Prediction of the Occupancy Rate of Future Time Intervals

The attempt to improve the prediction of the occupancy rate of a time interval in the near future by means of the number of purchased electronic tickets in previous time intervals is based on a linear model of the day curve of electronic tickets. The model uses the mean values $\overline{p}_t$ and the covariance matrix **S** of electronic tickets purchased in previous time intervals to compute the predictor

$$\widehat{p}_{t+i} = \overline{p}_{t+i} + \mathbf{S}_{t+i;t-j,\ldots,t}\mathbf{S}^{-1}_{t-j,\ldots,t;t-j,\ldots,t}\left(p_{t-j,\ldots,t} - \overline{p}_{t-j,\ldots,t}\right) \tag{5}$$

of the number of electronic tickets, which will be purchased in time interval *t* + *i*, given the number of electronic tickets purchased in time intervals *t* – *j*,...,*t*. The predictor of the occupancy rate in time interval *t* + *i* is again computed by a linear model, which uses the predictor of the number of electronic tickets $\widehat{p}_{t+i}$ as predictor:

$$\widehat{o}_{t+i} = \overline{o}_{t+i} + k\left(\widehat{p}_{t+i} - \overline{p}_{t+i}\right) + d \tag{6}$$

The model parameters $\mathbf{S}$, $\overline{p}_t$, $\overline{o}_t$, k and d are estimated from empirical data.

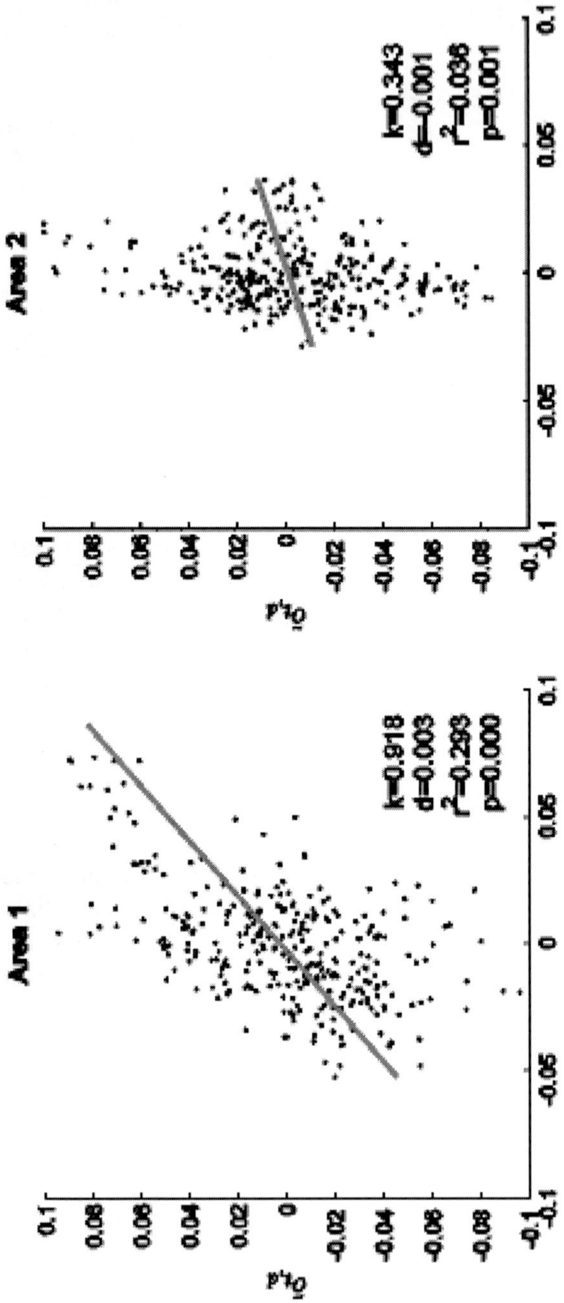

Figure 6. Linear regression of the residuals of occupancy rate o~t,do~t,d on the residuals of the number of mobile phone parkers p~t,dp~t,d in 1st district (area 1) and 6th district (area 2)

A separate model was developed for both areas and subject to three cross-validation rounds. Within each round, two of the three survey weeks served for calibration, while the remaining week served for validation. In the prediction model given by eq. (6) we chose $j = 0$, i.e., only the electronic tickets purchased in the current time interval were used for prediction. Figure 7 shows the Root Mean Square Errors (RMSE) of occupancy rate predictions calculated 0, 30, 60 and 90 min ahead. Each chart shows a separate cross-validation round, which corresponds to one of the three weeks. The red line indicates the RMSE of predictions based on the average day curve model $\widehat{o}_{t+i} = \overline{o}_{t+i}$. The real-time indicator could in most cases not reduce the prediction error of the average day curve model.

A direction for future work would be to examine the prediction accuracy of an implementation in a Kalman filtering framework with measurement model $p_t^* = ho_t^* + \epsilon_t$ and process model $o_{t+1}^* = \mathbf{F}_t o_t^* + \eta_{t+1}$ with observations $p_t^* = p_t - \overline{p}_t$, hidden state variables $o_t^* = (o_t - \overline{o}_t)$, observation model $h = k^{-1}$, observation noise $\epsilon_t \sim N(0, \sigma_t)$, state transition model $\mathbf{F}_t = \mathrm{cov}(o_{t+1}, o_t)/\mathrm{var}(o_t)$ and process noise $\eta_{t+1} \sim N(0, \xi_t)$.

7.3 Prediction of the Occupancy Rate in Exceptional Situations

Exceptional situations are by definition rare and will hardly come up numerously in three weeks survey period. We identified only one event, in which the occupancy rate differed considerably from the usual level. It was Shrove Tuesday in the 1st week, an event at which particularly many people visit the bars and pubs in the inner district of Vienna (area 1). This situation is included as one of three days in the upper left chart of Fig. 7. In this particular case it turns out that the real-time model predicts the occupancy rate clearly better than the average day curve model. The error improvement is strongest in the current time interval, but it still appears in predictions of future time intervals. This result indicates that the real-time information about electronic parking tickets might increase the accuracy of an occupancy model in situations, in which the demand of parking place deviates from the usual level.

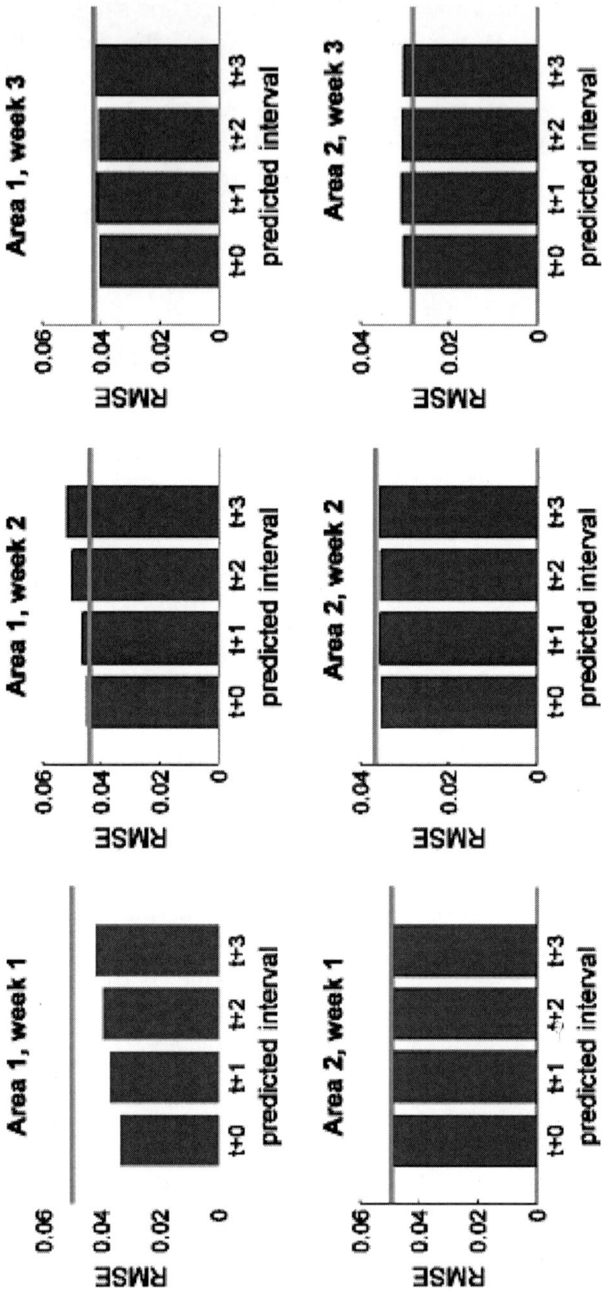

Figure 7. RMSE of a cross validation of occupancy rate predictions based on an average day curve model without (*red line*) and with (*blue bars*) real-time indication of purchased electronic tickets for time intervals t + 0 to t + 3

8 CONCLUSIONS AND OUTLOOK TO MODEL APPLICATION

We developed and tested a real-time occupancy model of short-term parking zones, which works without roadside infrastructure, but only uses existing real-time data sources. The following lessons we have learned from our work:

- Among the tested real-time data sources (electronic parking tickets, counts of car parks, traffic flow data) only the number and location of electronic tickets is able to indicate the occupancy rate in a given time interval and street section. However, this indicator is not yet available due to privacy concerns of the major mobile network operator in Austria.

- The occupancy rate of on-street parking spaces follows a strongly recurrent pattern on different days. An average day curve model thus predicts the occupancy at a given time interval very well. For the usual day-to-day variations of the occupancy rate it can hardly be outperformed by a real-time model.

- Unusual deviations from the day curve due to exceptional events are better predicted (and can only be predicted) by the real-time model. We had only one exceptional event in our three weeks survey, but such events will occur more often in the long run. There are various factors that significantly influence human behaviour: holidays, day characteristics such as working day, weekend, day before or after holiday or weekend, big events, etc. This makes such a model very useful. Its further development requires data from longer periods. It is not necessary to anticipate the exceptional events; they would simply appear in the real-time signal, so that the model can learn from its own history for future events.

The unavailability of the location data of electronic parking tickets is not necessarily the end of the story. The negative decision was based on an extremely narrow interpretation of the legal regulations, so that we see a chance for a positive decision in a second attempt. A significant improvement can also be expected from the new standards for Intelligent Transport Services (ITS) and their subsection Cooperative IT, which are currently developed by the standard organisations ISO, CEN, and ETSI. The new standards will start at 2016 with the availability of new cars. The solution will be a combination of GPS-data, vehicle to vehicle (V2V) and vehicle to infrastructure (V2I and I2V) communication. They will deliver location data in the range of five meters. It can be used as real-time indicator of parked vehicles with unreached accuracy (among many other things).

A simple option to make the information service available to the users would be a stand-alone webpage as shown in Fig. 8, where the user can chose the area and time for which the occupancy information shall be provided. The main goal is however to integrate the service into existing information services, e.g.

- the intermodal real-time traffic service AnachB.at;
- the mobile payment service HANDY Parken;
- the Viennese parking guidance system;
- traffic information services of broadcasting stations.

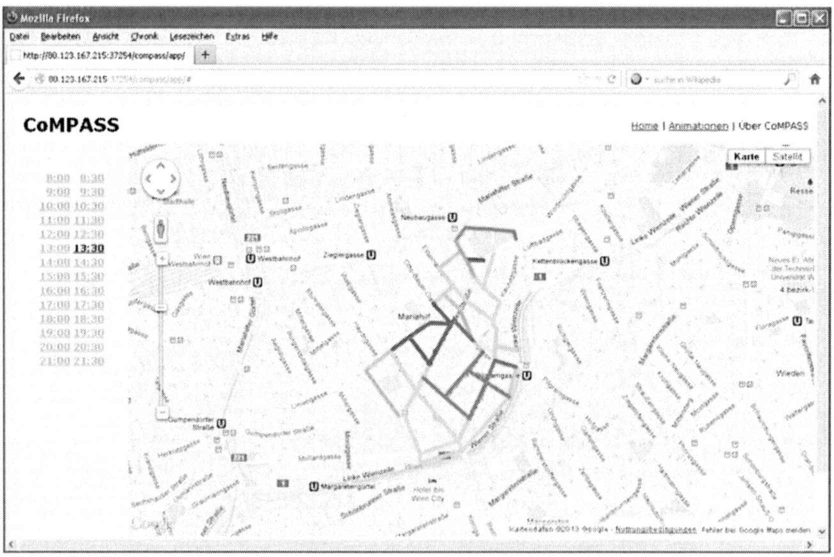

Figure 8. Internet information tool based on the real-time model

The end users' benefit of the information service is in either case a better knowledge of the current and forthcoming occupancy of on-street parking spaces at the destination. It could also include recommendations for alternative options in case of high parking pressure such as off-street parking facilities, intermodal transport, or public transport.

ACKNOWLEDGMENTS

Funding for this project is provided by the Federal Ministry for Transport, Innovation and Technology represented by the Austrian Research Promotion Agency (FFG).

REFERENCES

1. Ando, R., Mimura, Y.: A study to develop an information providing system on travel time. Int. J. Intell. Transp. Syst. Res. **8**(2), 77–84 (2010)

2. Basu, P., Little, T.D.C.: Wireless Ad Hoc Discovery of Parking Meters, Multimedia Communications Laboratory (MCL) Technical Report No. 01-08-2004 (2004)

3. Berenger Vianna, M.M., Da Silva Portugal, L., Balassiano, R.: Intelligent transportation systems and parking management: implementation potential in a Brazilian city. CITIES **21**(2), 137–148 (2004)

4. Cenerario, N., Delot, T., Ilarri, S.: A content-based dissemination protocol for VANETs: exploiting the encounter probability. IEEE Trans. Intell. Transp. Syst. **12**(3), 771–782 (2011)

5. Directive 95/46/EC of the European Parliament and of the Council of 24 October 1995 on the protection of individuals with regard to the processing of personal data and on the free movement of such data.

6. Dorner, A., Herry, M., Schuster, M.: Parkraumbewirtschaftung in Wien (Parking space management in Vienna). Werkstattbericht der Stadtplanung Wien, Magistratsabteilung 18, Vienna, Austria (1997)

7. Federal Highway Administration: Advanced Parking Management Systems: A Cross-Cutting Study. U.S. Department of Transportation, Washington (2007)

8. Gongjun, Y., Weiming, Y., Rawat, D.B., Olariu, S.: SmartParking: a secure and intelligent parking system. IEEE Intell. Transp. Syst. Mag. **3**(1), 18–30 (2011)

9. Hirahara, K., Ikeuchi, K.: *Detection of street-parking vehicles from panoramic street image*, Proceedings Intelligent Transportation Systems, IEEE, 12–15 October 2003, **2**:993–998 (2003)

10. Jaeseok, K., Haeyong K., Jongsoo, J., Yongbin, S., Pyeongsoo M.: Field Deployment of a Large-Scale WSN for Parking Management System, Sensor Mesh and Ad Hoc Communications and Networks (SECON), 7th Annual IEEE Communications Society Conference, 21–25 June 2010, pp. 1–3 (2010)

11. Käfer, A., Fürst, B., Pediaditis, E., Gabert, K.: Straßenverkehrszählung Wien 2010. Auswertung Gemeindestraße A + B (Road traffic census Vienna 2010. Analysis of communal roads A + B). Magistrat der Stadt Wien, Magistratsabteilung 18, Vienna, Austria (2001)

12. Markoff, J.: Can't Find a Parking Spot? Check Smartphone, Online: http://www.nytimes.com/2008/07/12/business/12newpark.html?_r=2&ref=business&oref=slogin&oref=slogin (10.01.2012) (2008)

13. Micheloni, C., Foresti, G.L., Snidaro, L.: A network of co-operative cameras for visual surveillance. Vision Image Signal Proc. **152**(2), 205–212 (2005)

14. Motor-Talk: Nie mehr Parkplatzsuche: Persönlichen Stellplatz online reservieren, (Never searching a parking space again: reserve a parking space online), Online: http://www.motor-talk.de/news/nie-mehr-parkplatzsuche-persoenlichen-stellplatz-online-reservieren-t18460.html (10.01.2012) (2002)

15. Orski, K.: Best space scenario. Traffic Technol. Int. **2003**, 54–56 (2003)

16. Pellecchia, A., Igel, C., Edelbrunner, J., Schoner, G.: Making driver modeling attractive. IEEE Intell. Syst.**20**(2), 8–12 (2005)

17. Riedel, R.: Parkraumbewirtschaftung in Amsterdam NL (Parking Space Management in Amsterdam NL), Stadtplanung Wien, Magistratsabteilung 18—Stadtentwicklung und Stadtplanung (2006)

18. Rodier, C., Shaheen, S., Eaken, A. M.: Transit-Based Smart Parking in the San Francisco Bay Area, California – Assessment of User Demand and Behavioral Effects. Transportation Research Record: Journal of the Transportation Research Board, No. 1927, pp. 167–173. Transportation Research Board of the Acadamies, Wanshington, D.C. (2005)

19. Rye, T., Hunton, K., Ison, S., Kocak, N.: The role of market research and consultation in developing parking policy. Transp. Policy **15**(6), 387–394 (2008)

20. Seong-eun, Y., Poh, K. C., Taehong, K., Jonggu, K., Daeyoung, K., Changsub, S., Kyungbok, S., Byungtae, J.: PGS: Parking Guidance System Based on Wireless Sensor Network, pp. 218–222. Wireless Pervasive Computing, ISWPC 2008, 3rd International Symposium, 7–9 May 2008 (2008).

21. Sifuentes, E., Casas, O., Pallas-Areny, R.: Wireless magnetic sensor node for vehicle detection with optical wake-up. IEEE Sensors J **11**(8), 1669–1676 (2011)

22. Sofortparken.at: http://www.sofortparken.at; retrieved at March 08 2013 (2013)

23. Tang, V.W.S., Zheng. Y., Cao, J.: An Intelligent Car Park Management System Based on Wireless Sensor Networks, pp. 65–70. Pervasive Computing and Applications, 1st International Symposium, 3–5 August 2006 (2006)

24. Unterluggauer, A.: AnachB - die neue Live-Verkehrsinformation der Vienna Region (AnachB - The New Live Travel Information in the Vienna Region). Paper presented at the AGIT 2009 Symposium und Fachmesse für Angewandte Geoinformatik, Salzburg, Austria (2009)

Chapter 3

Analysis of Intelligent Transportation Systems Using Model-Driven Simulations

Alberto Fernández-Isabel and Rubén Fuentes-Fernández*

Departamento de Ingeniería del Software e Inteligencia Artificial, Facultad de Informática, Universidad Complutense de Madrid, 28040 Madrid, Spain

ABSTRACT

Intelligent Transportation Systems (ITSs) integrate information, sensor, control, and communication technologies to provide transport related services. Their users range from everyday commuters to policy makers and urban planners. Given the complexity of these systems and their environment, their study in real settings is frequently unfeasible. Simulations help to address this problem, but present their own issues: there can be unintended mistakes in the transition from models to code; their platforms frequently bias modeling; and it is difficult to compare works that use different models and tools. In order to overcome these problems, this paper proposes a framework for a model-driven development of these simulations. It is based on a specific modeling language that supports the integrated specification of the multiple facets of an ITS: people, their vehicles, and the external environment; and a network of sensors and actuators conveniently arranged and distributed that operates over them. The framework works with a model editor to generate specifications compliant with that language, and a code generator to produce code from them using platform specifications. There are also guidelines to help researchers in the application of this infrastructure. A case study on advanced management of traffic lights with cameras illustrates its use.

KEYWORDS

intelligent transportation system; smart city; sensor; actuator; traffic lights; simulation; model-driven engineering; modeling language; agent-based modeling; code generation

1. INTRODUCTION

The concept of *smart city* appears as an answer to the challenges posed by the management of highly interconnected physical and Information Technology (IT) infrastructures in relationship with their communities and environment [1]. The goal is to take advantage of the collective intelligence in such complex systems to get more efficient, sustainable, and liveable environments [2]. Intelligent (or Smart) Transportation Systems (ITSs) [3] play a key role in this context. Their services try to improve traffic regarding security, sustainability, and reduced costs in terms of time, money, energy, and environmental impact. Examples of these services [4,5] are the timely update of traffic data, the request of the equipment needed to assist in emergencies using the data provided by vehicles, or the tracking of freight along its journey. For this purpose, these systems make use of the pervasive presence of networks of sensors and actuators (Sensor Networks, SNs) deployed in people, vehicles, and the environment of *smart places*, and the high availability of data regarding participants in traffic and their activities [6].

The development of these complex systems faces important challenges [3,5,6]. First, their potential effects on the environment and living beings make it difficult to set up suitable controlled environments for experiments. Second, these systems usually comprehend multiple distributed components (both hardware and software), frequently embedded in other elements such as vehicles, phones, traffic lights, or roads. This kind of system has high costs of development, testing, and deployment. Third, there are many interacting elements in a real traffic setting, what makes difficult to validate hypotheses on the behavior of ITSs.

Simulation appears as a useful tool to deal with the previous issues [7]. It allows controlling the relevant variables of the problem, running multiple experiments, and the incremental development of ITSs (*i.e.*, gradually replacing models and software implementations by the actual components and environment). However, this approach also presents some drawbacks [8], mainly concerned with the difficulties to guarantee the alignment between the resulting simulation and the initial abstract model.

Model-Driven Engineering (MDE) [9] has been proposed as a way to overcome these limitations of simulations [10]. In a model-driven development, researchers mainly specify their simulations using models compliant with well-defined Modeling Languages (MLs). These models are combined and refined (adding new information or modifying it) until it is possible to generate code from them. Transformations automate some recurrent modifications of models, code, and other related artifacts. In this way, all the information required to produce the simulation is explicitly represented. This facilitates validating artifacts with theories and crosschecking among experiments. For instance, the same theoretical model of drivers' behavior can be the basis for simulations in different platforms, different models of traffic flows in a city can be compared with implementations in the same platform, and all of them can be reused and validated in different projects as long as they keep the same theoretical basis or simulation infrastructure.

This paper presents a MDE framework to develop simulations of ITSs. This framework comprehends: a ML for ITS experts; a *model editor* and a *code generator* tools; and *guidelines* to apply these elements. Figure 1 provides an overview of the framework components and the main related works that influence their development. All considered frameworks adopt MDE approaches, and all their MLs follow the agent paradigm [8,11].

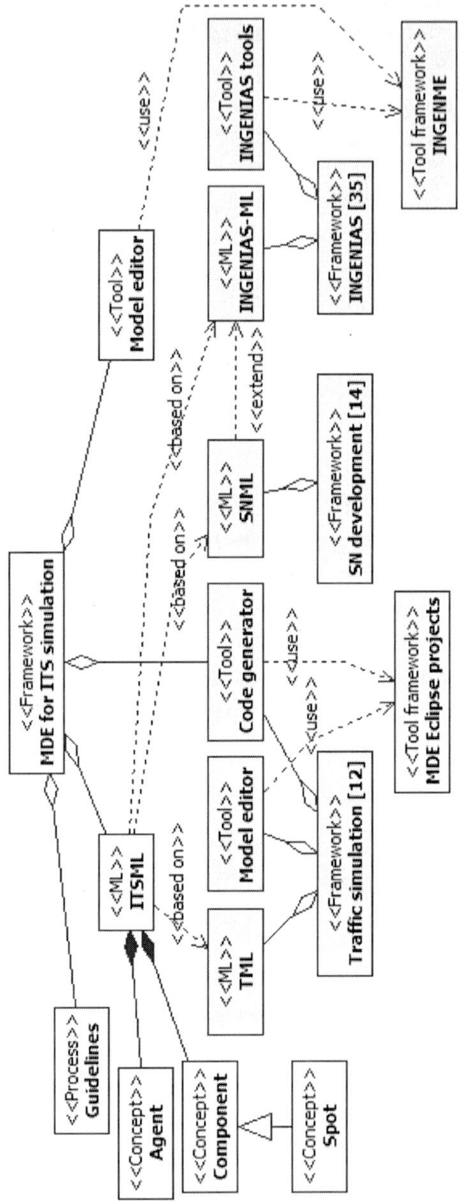

Figure 1. Main components of the Model-Driven Engineering (MDE) framework for Intelligent Transportation System (ITS) simulation (in white) and related works (in yellow). Aggregation relationships (with diamonds) represent whole-part relations, and dependency relationships (discontinuous lines) different relations of use.

Our ITS ML (ITSML) builds on the foundation of three MLs that address relevant aspects of the problem: the simulation of traffic in [12] (the Traffic ML, TML), and the development of SNs in [13] (the SN ML, SNML), which in turn is partly based on the INGENIAS methodology [14] for agent-oriented software development and its INGENIAS ML (the INGENIAS-ML). The ITSML provides a seamless integration of their concepts (including direct reuse of parts of their MLs) and additional elements. The integration is based on the agent paradigm underlying these languages. Thus, the ITSML distinguishes between *agents* (*i.e.*, intentional and social entities) and *components* (*i.e.*, elements of systems that are not agents).

Components have a state, are able to produce events, and make available interfaces with methods to manipulate them. Most concepts from the other MLs become *components* here, e.g., the devices of SNs and the elements of the environment in traffic. The execution of methods can be linked in workflows through the production and use of information (e.g., external notifications or internal facts). This ML also introduces the concept of *spot*, as a *component* that indicates where devices can be placed and what they can perceive and act on. For instance, persons, vehicles, and traffic lights are *spots*.

Agents represent both people (e.g., *person* in [12]) and software agents (e.g., *managers* of devices in [13]). They follow a perceive–reflect–act cycle, with specific *components* for each of these tasks. This facilitates decoupling the interaction with the environment and the modification of their internal state.

As general features, the ITSML provides inheritance and instantiation mechanisms. The first one supports the specialization of concept types at the model level. The second one allows defining instances of any type in models, which makes possible the full specification of the simulation at that level. There is also a general association relationship between concepts that can be stereotyped.

The tools of the framework are a model editor developed using the meta-editor INGENME [15], and the code generator from [12]. The first one supports the specification of models compliant with the proposed ML. The second one allows mapping elements in models to code templates that are used to generate the source code of simulations.

The case study that illustrates the use of this infrastructure is the simulation of the management of traffic lights in [16]. That work uses the

Simulation of Urban MObility (SUMO) open traffic simulation suite. The paper considers traffic lights that optimize flows in junctions by perceiving the length of their queues and coordinating among them. This case study addresses the development of such simulation using the proposed approach. It shows how most of the simulation can be specified with models and then to generate code from them, which reduces coding effort and facilitates its understanding and later modification. Additional experimentation is also discussed more briefly.

The rest of the paper is organized as follows. Section 2 reviews the background of this work in MDE (see Section 2.1), and the development frameworks for traffic simulations (see Section 2.2) and SNs (see Section 2.3). The framework for the model-driven development of ITS simulations (see Section 3) is described through its ML (see Section 3.1) and its development guidelines and tools (see Section 3.2). Section 4 applies it to the case study of the simulation of intelligent traffic lights (see Section 4.1) and presents additional experimentation (see Section 4.2). The results are compared with related work in Section 5. Finally, Section 6 discusses some conclusions and future work.

2. BACKGROUND

The approach for the development of simulations of ITSs is based on four main works, which are discussed in the following subsections. MDE (see Section 2.1) is the general paradigm for software development used in this research. It is focused on models and their semi-automated transformations. Three development frameworks adopting this paradigm contribute to our research: those for traffic simulations [12] (see Section 2.2) and SNs [13], this extending the INGENIAS methodology [14] (see Section 2.3). The proposed framework integrates elements from these works with specific extensions for ITSs, both at the levels of ML and tools.

2.1. MDE

MDE [9] organizes software development around *models*. These gather most of the information required to produce systems and their related artifacts. The development process is mainly conceived as an iterative refinement of models, where information is added to them and modified.*Transformations* automate some of these modifications, and also generate from models artifacts such as documentation, code, or tests. Examples of transformations can be mapping abstract theoretical models to design ones for a given simulation platform, or adding a default goal of path-following to a driver agent. There are alternatives to work with these elements.

In order to enable the automated processing of models by tools, MDE needs to define formally their MLs. Metamodels are the most widely accepted means for this task regarding graphical graph-oriented MLs, which are the most used [9]. A metamodel defines the primitives available in the language, as well as the constraints applicable to them.

Metamodels are specified using meta-modeling languages. The Meta-Object Facility (MOF) [17] is the standard defined by the Object-Management Group (OMG). It is used to define popular standard MLs, such as the Unified-Modeling Language (UML) [18]. However, tool support for MOF is limited, so researchers and engineers frequently resort to better supported alternatives. Ecore [19], from the Eclipse modeling projects, has extensive tool support and is aligned with the subset of MOF known as Essential MOF (EMOF) [20]. MOF-based MLs use *references* to relate concepts. These references are directed binary links with only a few properties (e.g., super-type and containment). This makes them cumbersome sometimes to specify complex relationships with attributes. The Graph-Object-Property-Role-Relationship (GOPRR) [21] meta-modeling language considers a richer set of primitives for this task. Its languages are organized around *graphs* that contain *objects* connected by *relationships*. Relationship ends are *roles*, and they can appear in any number. All the previous elements can have *properties*. Some tools based on GOPRR are Metaedit+ [21] and INGENME [15].

MDE approaches process information automatically using transformations. Most times, their inputs and outputs are models compliant with metamodels, or text-based artifacts (e.g., code or documentation) compliant with grammars [22]. There are multiple ways of implementing these transformations [23], from modules written in mainstream imperative programming languages, to mappings written in specific declarative transformation languages and executed by engines. Examples of the first approach are wizards in [12] and modules in INGENME [15]; JET [24] and ATL [25] are popular Eclipse-supported examples of the second approach.

The organization of the overall development process also has multiple alternatives. One of the best known is the Model-Driven Architecture (MDA) [26] of OMG. It defines three levels of abstraction for its MLs. Computation Independent Models (CIMs) are intended for domain-specific information, here traffic and very abstract views of ITSs. Platform Independent Models (PIMs) describe general computational abstractions, not depending on specific architectures, e.g., sensors, actuators, and controllers of ITSs. Platform Specific Models (PSMs) include technology-related information, e.g., the target simulation platform. INGENIAS proposes two MDE processes, one based on the Unified Process [27] and another on Scrum [28]. They are oriented to different types of project and team. Both use almost the same kinds of diagram over the entire development, independently of the level of abstraction of the stage. In all the considered processes, designers manually modify development artifacts or run automated transformations on them.

Transformations can also be used to translate information among models of MLs compliant with different meta-modeling languages [29]. This is a less frequent type of transformation, and its specification is quite complex.

2.2. The Traffic Simulation Framework

The *traffic simulation framework* [12] proposes a MDE approach to develop this kind of simulation. It comprehends the TML to describe theories of traffic, and tools to work with it.

The TML is based on Agent-Based Modeling (ABM) [8]. Its agents are intentional entities that represent *persons*. They are modeled in terms of the *goals* they pursue and the *tasks* they can execute to achieve the former. Internally, these agents follow a perceive–reflect–act cycle [11]. An agent perceives its external environment, uses this information to update its internal state, and then to select a suitable goal and a related task according to it, executes this task, and starts again the cycle. This cycle is modeled using *evaluators* (which perform all these activities but the execution of tasks), and *actuators* (that execute the selected tasks).

Regarding traffic, the TML adopts the Driver–Vehicle–Environment (DVE) [30] model. This considers the dynamic interactions among *drivers* moving in an *environment* using their *vehicles*. In these interactions, the mutual influences among the different components affect their features and behaviors. For instance, the mental workload of a driver largely depends on the weather conditions, which are part of the environment. The TML modifies this basic model to be able to integrate different theories of traffic. Its participants are *persons* (that are also the agents), who can drive *vehicles* or not. *Persons* are characterized in terms of the *knowledge* they have and the features of their *profile*.

The ML uses *methods* to specify the actual effects of the interactions among the previous elements. Methods act as placeholders where indications can be provided on how to update dynamically features. There are default methods for several types of entity. An alternative to model these effects is to use the inheritance mechanisms to provide new types of relationship (e.g.,*affect* or *impede*), but this can be difficult to specify and semantics not being clear.

Most of the previous concepts can be decomposed in parts (*i.e., components*). For instance, *Profile*has *PComponents*, and *Environment EComponents*. They can also be extended through inheritance.

The implementation of this framework is based on Eclipse projects. An Ecore metamodel defines the TML, and related tools are based on the Eclipse Modeling Framework (EMF) [19] and the Graphical Editing Framework (GEF) [31]. The two main tools are a model editor and a code generator.

The model editor allows generating graphically instances of the metamodel, *i.e.*, models. These are drawn as graphs of the concepts and relationships included in the metamodel. This editor is directly generated from the metamodel using the EMF and GEF functionality.

The code generator integrates information from different sources to output the source code of simulations. Its inputs include, among others: the traffic metamodel; models compliant with it; the classes EMF generates from the metamodel; libraries describing the target simulation platforms; code templates for specific parts of the generated code; and mappings from model entities to those templates. A template includes text and placeholders that indicate where and what information from models needs to be inserted. The generator manages all this information with graphical wizards. It is implemented using Java with additional libraries.

2.3. The SN Development Framework

The *SN development framework* [13] introduces a MDE approach to develop SNs. It is based on the INGENIAS methodology [14] and its tools for model editing [15], and Eclipse transformations [24,25].

INGENIAS is a MDE methodology to develop Multi-Agent Systems (MASs). MASs are a general paradigm (not only for simulations) where the basic abstraction is the *agent*. As it happens in ABM, INGENIAS agents are intentional and social entities. They execute *tasks* pursuing certain *goals*. These *tasks* can change the internal state of the *agent* or its external environment, or trigger interactions. The environment is modeled using *environment applications*, which produce *events* and have *methods* to act on them. Sometimes, these *events* trigger *tasks*. Interactions exchange information among *agents*.

The INGENIAS-ML supports limited forms of inheritance and instantiation relationships for agents. The latter defines the actual *agent* instances that exist in the system from their types. This is used, for example, to specify the system deployment.

The INGENIAS tools are currently based on the INGENME framework [15] for MDE. INGENME supports the definition of MLs based on its implementation of GOPRR [21]. Its meta-editor allows generating functioning model editors from metamodels. These editors are similar to those of GEF [31], but simpler to describe, as they involve less intermediate models. Additional modules can be integrated in the basic version of these editors. These modules are implemented in Java, and have access to programming interfaces provided by the framework to manipulate both metamodels and their models. Among other tasks, they implement transformations. Text-generation modules are based on templates, as in the traffic simulation framework (see Section 2.2).

The SN development framework extends the INGENIAS-ML to model SNs with the SNML. Its models are organized around *containers*. *Containers* are computational devices where other*resources* run. *Resources* can be pure software *utilities* or *devices* with hardware components (*i.e.,sensors* and *actuators*). Communication *channels* linking *containers* are considered a particular type of *resource*. The control of these *resources* is the responsibility of *managers*, which is a role that *actor*agents play. These *managers* can be *passive* or *active controllers*. The first group only listens to notifications from *resources*, while the second group can also invoke their modification methods.

The framework includes basic guidelines on how to model with the SNML and translate these specifications to the INGENIAS-ML. The rest of its development follows an INGENIAS process [27].

3. THE MDE FRAMEWORK FOR ITS SIMULATION

The framework for the development of ITS simulations comprehends the following elements: a ML (see Section 3.1), and guidelines to apply it with a model editor and a code generator (seeSection 3.2).

3.1. Modeling Language

The ITSML is organized around the main concepts that appear in this kind of system and its context (see Figure 2). There are *persons* who interact with *vehicles* and an external *environment*, which is composed of *things*. All these elements are *places* that can contain multiple *spots*. A *spot* is an element that can be observed and/or where *containers* can be located. *Containers* are computational nodes, linked by communication *channels*. In them, software *managers* control the deployed *sensors* and *actuators*. All the previous concepts share some features (e.g., they are modeled with a state and an access interface with methods, and generate events). A common abstract super-concept named *component* gathers these features.

That conceptual basis is the result of integrating and extending multiple researches related to the domain. The ML follows an ABM approach [8], so it makes of agents, and their interactions among them and with external objects, the core of modeling. This perspective is common to several MLs that provide concepts for the ITSML. The TML [12] is the main one, given the close relationship between the domains of traffic and ITS simulations. The INGENIAS-ML [14] offers a rich set of primitives for ABM and basic extension mechanisms of limited inheritance and composition. It is also the foundation of the SNML [13], which models concepts of devices and their controllers in SNs.

The ITSML specifies that conceptual framework using a metamodel. Working with metamodels allows reusing parts of the other MLs. For instance, it supports introducing and modifying concepts and relationships as specializations of those present in the different base metamodels. There are also extensions that mechanisms in the base MLs do not support, but those of metamodels do. For instance, the ITSML has to allow defining instances of arbitrary types in its models.

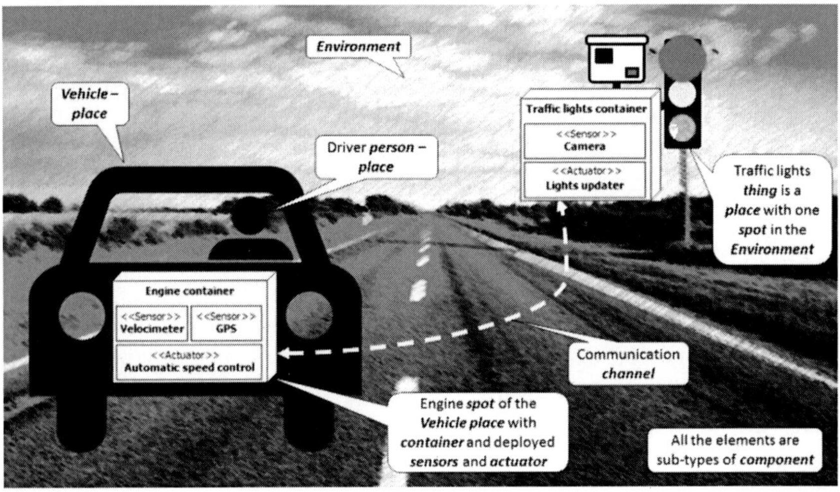

Figure 2. Main concepts (in *italics*) of the ITS Modeling Language (ITSML) in context.

The meta-modeling language used for the ITSML is the implementation of GOPRR [21] in the INGENME framework [15]. In next diagrams of this section, nodes and relationships, respectively, represent meta-entities and meta-relationships. The ends of these meta-relationships are the meta-roles. Meta-properties appear as attributes or adornments of the previous elements. Examples of these are the numbers in the meta-roles that are cardinality indications. Meta-relationships with triangles represent inheritance, and those with diamonds aggregation, with filled diamonds representing composition and hollow diamonds mere aggregation. This notation is similar to that of UML [18]. As the ITS metamodel integrates concepts from different metamodels, these have been located in packages. Names of concepts and relationships from other metamodels appear qualified with the name of its package, *i.e.*, *traffic* [12], *SN* [13], or *INGENIAS* [14].

The rest of the section discusses in detail the concepts for ITSs in the metamodel. It starts with the general *component* (see Section 3.1.1). Then, it focuses on the *places* where *sensors* and *actuators*can be deployed and with which they interact (see Section 3.1.2). *Managers* control these resources to provide services to *persons* (see Section 3.1.3) moving in an *environment* (see Section 3.1.4). There are also some general mechanisms applicable to all these concepts (see Section 3.1.5).

3.1.1. The *Component* Entity

Figure 3 is focused on the *component*, which is the base concept in most of the inheritance hierarchies of the ITS metamodel. The SN *device* and the constituents of the traffic *person* (with the exception of the mental entities *goal* and *knowledge*), *vehicle*, and *environment*, become *components*. *Components* have an internal state and an interface composed of *methods*. *Methods* can be *external*(when they can be invoked from outside their *components*), or *internal* otherwise.

The state and methods are characterized in terms of information. Pieces of *information* can be of three types: *facts* make up the internal state; *events* are information that appear in *components* but they do not explicitly produce, and that can be used internally or externally; *notifications* are information produced for external use. *Methods* output both *facts* and *notifications*, but *events* come from non-specified parts of *components*. Besides their results, methods are defined by the parameters they need and the conditions that enable them, which can involve any kind of*information*. *Components* can be connected in pipelines using the *information* their *methods* produce and consume.

The metamodel defines two convenience meta-relationships for *methods* and their related information: *triggers* and *perceives*. The first one relates a *component* to a *notification* that one of its*methods* generates. The second one links a *component* to any *information* that is input to one of its*methods*. These meta-relationships allow focusing on the considered information and omitting the particular *methods* involved.

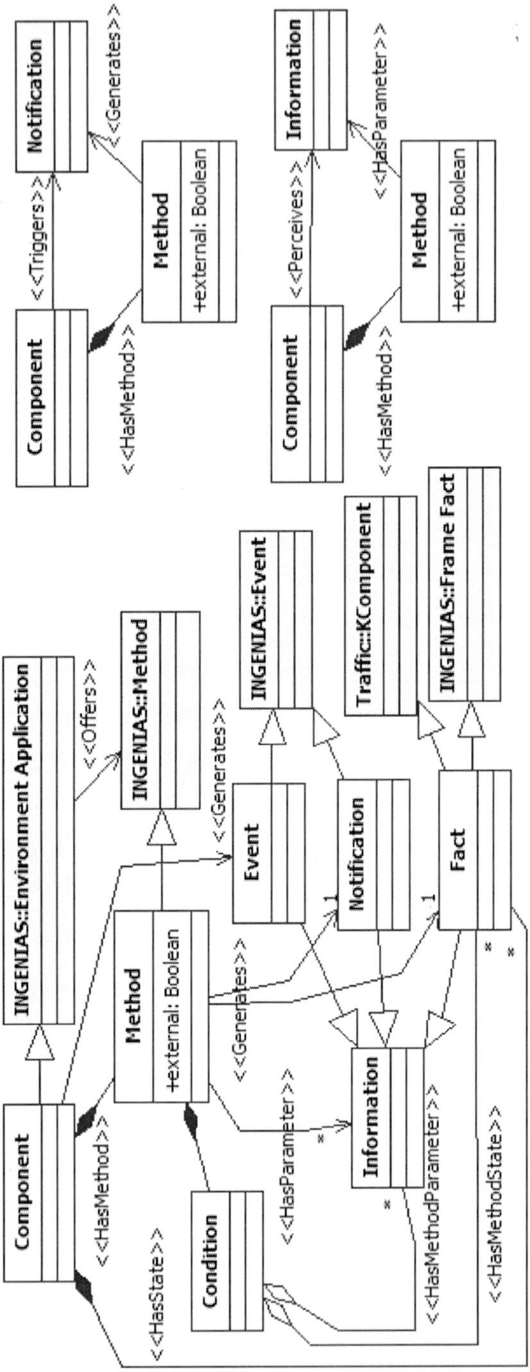

Figure 3. Partial ITS metamodel. *Component* related elements.

3.1.2. *Place* and *Device* Related Entities

An ITS needs to gather information from its surroundings and act on them. Figure 4 shows the high-level structure of this context. It is abstracted as a set of *places* that contain multiple *spots*. These *spots* are the *components* that can be actually monitored and modified, and where devices can be placed. Following the DVE model adopted by the TML (see Section 2.2), the participants in the traffic phenomena (*i.e., persons*, their *vehicles*, and the *things* in the external *environment*) are *places*. Their components according to the TML become *spots*.

Following the SNML (see Section 2.3), *devices* are responsible of sensing and acting in the ITS surroundings. The metamodel incorporates the device-related concepts from SNs as *components*. A*container* is the computational resource physically attached to a *spot*. Communication *channels* link them. A *container* runs multiple *devices* (*i.e., sensors, actuators,* and *utilities*). Both *sensors* and*actuators* can invoke consult *methods* of *spots*. *Sensors* can also perceive their *notifications*, and*actuators* invoke their external modification *methods*. *Utilities* are pure software components without access to the surroundings.

Sensors and *actuators* are thus related to two *spots*, which can be (or not) different: the one to which their *container* is attached (*i.e., HasAttached* meta-relationship); and the one with which they interact. For instance, a *place vehicle* has a *spot* for a *container* with a *thermometer sensor* that measures the *air* temperature, being that *air* a *spot* in the *place weather conditions* of the *environment*. The *perceives* and *actuates* meta-relationships are convenience meta-relationships for these interactions. They link *sensors* and *actuators* with the *spots* they manipulate without considering the involved *methods*.

The control of *devices* can be modeled in their *methods* or using *manager agents*. The activation of a*method* happens when its inputs (e.g., external *notifications* and internal *events* or *facts*) and the internal state of its *device* meet some *conditions*. As a consequence, it generates new information, either internal to the *device* (as *facts*) or external for other components (as *notifications*). For instance, a *timer sensor* in a traffic light has a *method* activated by a *time-consumed event*. When this happens, this *method* produces a *change notification* for the *actuator* in charge of changing lights, which processes it with another *method*. This *method* changes to a target light that depends on a *fact* that represents the current light, and which is part of the internal state of that actuator. On the other hand, *manager agents* (see more on them in Section 3.1.3) control *devices* when complex algorithms are needed (as they have a rich model for decision-making) or there is communication among*containers* (which is one of the specific responsibilities of agents).

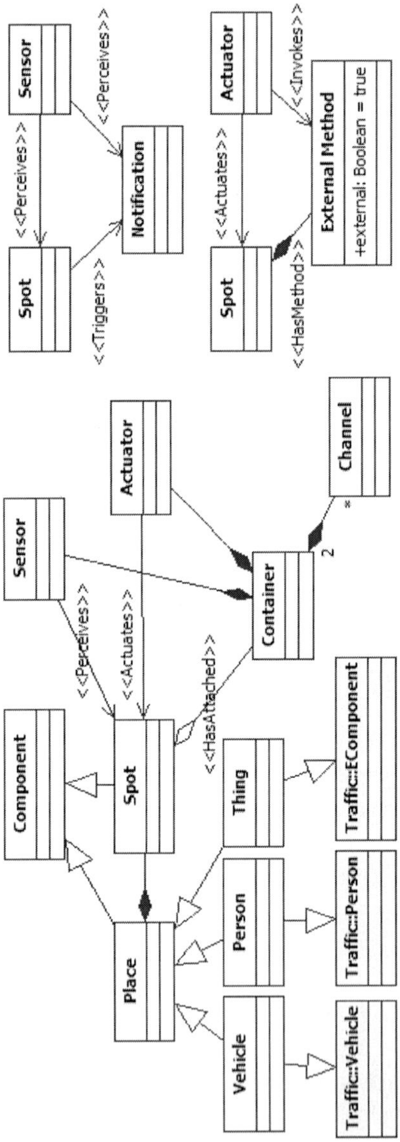

Figure 4. Partial ITS metamodel. *Place* and *Device* related elements.

3.1.3. The *Person* and *Manager* Entities

The ITS metamodel models with *agents* complex intentional and social components. These include *persons* and *managers* that control *devices* (see Figure 5). The specification of *agents* includes the information they have, their goals and capabilities, and the process to manage the previous elements.

Knowledge specifies the internal state. It includes *facts* (*i.e.*, any statement about the world) and*goals* (*i.e.*, states of the world that an *agent* wants to keep or to achieve).

Tasks are the high-level actions of agents. They are implemented using *methods* or other *tasks*. When using *methods*, a *task* is eligible for execution depending on the *condition* of the *method*. In this case, the *task* can change the *agent*'s *knowledge*, use *devices*, or act on the environment through*notifications*. Communication among agents is modeled as a particular use of *notifications*. The alternative to define the execution of a *task* is as a workflow of interconnected *tasks* to set up complex activities. These workflows follow the INGENIAS-ML [14].

Agents work in a perceive–reflect–act cycle that defines its dynamic behavior. This cycle is modeled with three types of *components* (*i.e.*, *sense*, *reflection*, and *acting*) that operate over their own and their *agent* internal state. The *sense* meta-class models perception as an interaction with observed *devices* that adds or modifies *facts* in the state of the *agent* or the *sense*. The instances of the*reflection* meta-class are responsible of the reflection stage. They analyze available *knowledge* (both*facts* and *goals*), and produce new one or modify the existing. Then, they check non-satisfied *goals*, and identify the *tasks* related to them and suitable for execution in the current context. Finally, the*acting* meta-class represents the actual execution of one of the executable *tasks*.

The *agent* concept is suitable to represent directly control *managers*. For *persons* involved in traffic, there are some modifications. A *person* has specific *person sense* and *person acting components*that allow interacting directly with *spots* without the need of intermediate *devices*.

The previous definitions allow aligning concepts from the ITSML with those of its base languages. The ITSML *agent*, *fact*, and *goal* are sub-types of the INGENIAS-ML meta-classes *agent*,*frame fact*, and *goal*. Both the TML *knowledge* and *profile* are modeled as ITSML *knowledge*, and the

TML *goal* is a super-type of the ITSML *goal*. The *person* sub-type of *agent* is also a subtype of the TML *person*.

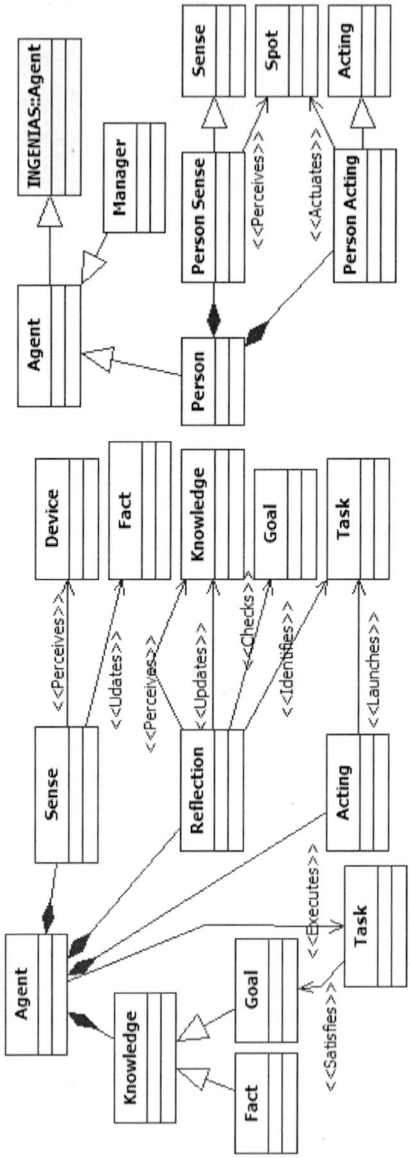

Figure 5. Partial ITS metamodel. *Person* and *Manager* related elements.

3.1.4. The *Environment* Entity

Regarding the *environment*, it represents the elements participating in traffic that are not people, their vehicles, or the ITS components (e.g., *devices* and their *managers*). Examples of its elements are the roads, the traffic signals, the weather conditions, or obstacles. It is an extension of the TML*environment* (see Figure 6). It is modeled as an aggregation of *things*, which are located over a *map*.

The *map* represents the distribution of the physical space. It is a kind of graph composed by a set of *section*, each one linking two *junctions* (at least one) and having a given *length*. This length can be measured in different units, e.g., kilometers or number of vehicle along it. A *section* is a set of *lanes*that can be direct (from the origin to the destination junction) or reverse (the other way). A *lane* has an *occupants* list of references to the vehicles or persons in it ordered according to their position.

The other elements of the environment (the equivalent of the TML *EComponent*) are *things*. They can be placed in specific *map locations*, but also being considered global for the environment, e.g., the weather conditions or the sun position.

Vehicles and *persons*, being *places*, can have a location. In the case of instances of *vehicle*, this is mandatory. *Persons* can have a location, when they are pedestrians, or being attached to a *vehicle*.

3.1.5. General Features

The ITS metamodel also includes general instantiation, inheritance, and association mechanisms. For each meta-class of name *name*, there is another meta-class of name *Iname* that represents its instances at the model level. An instance of the later can be linked to an instance of the former with an instance of the meta-relationship *instanceOf*. For example, there is a meta-class *IPerson* linked by a meta-relationship *instanceOf* to the meta-class *Person*, so in a model, a specific driver *IPerson* can be defined as an *instance of* a specific type of *Person*. The inheritance mechanism extends those available in the traffic, SN, and INGENIAS MLs. It can be applied to most of concepts in the ITS metamodel, with the semantics that the sub-concept inheriting from the super-concept has all its features, but it can extend or constrain them with additional elements.

Default constraints establish that an animated concept (*i.e.*, an instance of the meta-class *person*) cannot inherit from a non-animated concept (*i.e.*, an instance of a meta-class that is not *person*), and *vice versa*. Regarding association, there is a general meta-relationship *relatedTo* between instances of *component*. It has a*stereotype* property to establish arbitrary types of it.

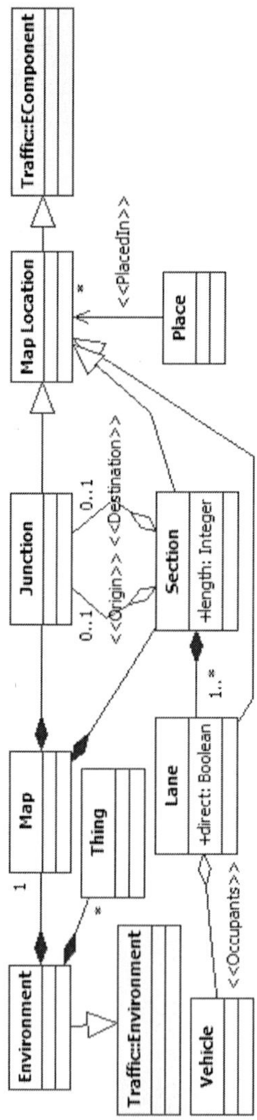

Figure 6. Partial ITS metamodel. *Environment* related elements.

3.2. Development Guidelines and Tools

Figure 7 shows the development guideline for the framework. The process has two different stages. The first one (see nodes 1–20) is specific of this work and focused on modeling with the ITSML. The definition of its metamodel (see Section 3.1) provides hints to identify the different concepts. The results of this stage are models that provide an abstract representation of the ITS and its environment. They are independent of specific target simulation platforms. The second stage takes these models as input and runs transformations to get models with the INGENIAS-ML (see activity 21), and then applies an INGENIAS development processes [27,28] to get the simulation code (see activity 22).

The process starts identifying *places* in decision 1. These establish the context of the system and the elements over which the ITS must work. If there is any *place*, activity 2 assigns it a name and identifies its *spots*. Next activities in this path specify the *place* as a *component* (starting with decision 3) and its *spots* (see decision 14). If there are not more *places* to identify, the abstract modeling of the ITS has finished, and the process starts the INGENIAS development (see activities 21 and 22).

The description of a *place* depends on its type (see decision 3): *thing, vehicle,* or *person*. All of them have to be specified as *components* (see activities 4–6). This specification indicates the information available from these elements, and the means to interact directly with them. Instances of *person*should not usually include any external *method* to modify its internal knowledge, as it would violate their autonomy according to the agent paradigm. Other of their features can be modified from the environment, for instance to reflect that they have suffered some injury. If needed, these*places* must also be located in the *environment map* (see activity 13), which also implies specifying this.

The specification of a *person* also comprehends its behavioral cycle. This is specified in terms of its *sense* (see activity 10), *reflection* (see activity 11), and *acting* (see activity 12) *components*. *Sense*instances use their *methods* to get *notifications* from external sources (*i.e.*, *spots* and *devices*), and include it as *facts* that are part of the internal state of the *person*. *Reflection* instances model several activities: generation of new derived *knowledge* from that already available; checking *goal*satisfaction according to the current internal state; selection of potential *tasks* to execute according to the state of *goals* and other *knowledge*. *Acting* instances represent the actual execution of *tasks*. Activity 7 is used to specify those *goals*, activity 8 the *facts*, and activity 9 the *tasks*. These elements are linked among them: *tasks* are potentially able to satisfy some *goals*; *goals* are satisfied according to conditions specified in terms of *knowledge*; and *tasks* are eligible depending on the state of associated *goals*.

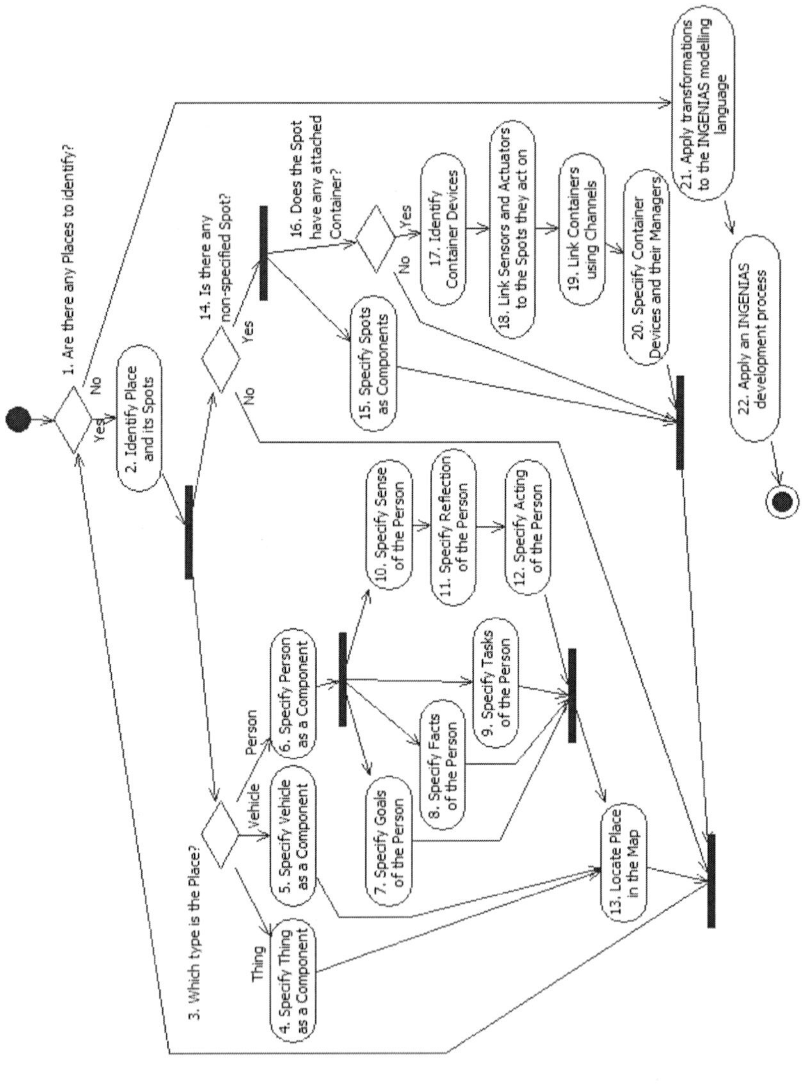

Figure 7. Activity diagram of the development process with the ITSML. Concepts starting with uppercase belong to this ML.

After specifying a *place*, the process continues with its *spots*. These *spots* are elements where*devices* can be located, or that *devices* can observe and act on. The specification of *spots* starts identifying them (see decision 14). Every *spot* must be defined as a *component* to determine its related *information* and *methods* (see activity 15). *Spots* are also the elements that can have attached*containers* (see decision 16). Activity 17 identifies the *containers* attached to a *spot* and the *devices*running in it. *Containers* are connected through the *channels* identified in activity 19. Among *devices,sensors* observe *spots* and *actuators* act on them. These *spots* can be or not the same to which their*containers* are attached. Activity 18 links *sensors* and *actuators* to their target *spot*. Finally, activity 20 specifies the *container devices* as *components* and their *managers* as *agents*, if they exist. Although not included in the figure, the specification of *managers* should follow a similar workflow to that of*person* (see activities 6–12), as both entities are sub-types of *agent*.

Activities 21 and 22 correspond to the transition and the development with an INGENIAS process [27,28]. The first activity develops, if not available, and runs an automated transformation with input ITSML specifications and output INGENIAS-ML models. Then it follows an INGENIAS development process to generate the simulation source code. The main issue here has to deal with code generation. It relies on platform-specific templates (see Section 2.3), which must be available from other projects or need to be developed from scratch.

This approach is in line with MDA [26]. The ITSML models are CIMs and PIMs, and the INGENIAS-ML models are PIMs and PSMs, in both cases depending on their level of abstraction.

Figure 7 shows only a sequential perspective of the process. In fact, its actual application needs to be iterative in its different paths. For instance, in the path from nodes 3 to 13, activities 4–6 can be done in the first iterations to give a high-level description of the system context, while the other activities correspond to a later and more detailed specification. In that path also, when modeling the *person tasks* (see activity 9), it is possible to find out that their execution requires additional internal *methods* of *person*, so activity 6 must be revisited. Moreover, when in activity 18 designers link *sensors* and *actuators* to the *spots* they work on, this can lead to modifications in the definition of those *spots* (made at some moment in activity 15), or even of their *places* (starting in decision 3).

The previous process is supported by two main tools. A graphical model editor allows generating model instances compliant with the ITS metamodel. This editor has been developed using the infrastructure of INGENME [15]. A code generator takes as input the metamodel and the model instances to generate the source code of the simulation for a target platform. Researchers only need to provide the code templates for that platform, and give the correspondences between those templates and the concepts of the ITS metamodel. The code generator has been developed for the traffic simulation framework [12]. It replaces here the modules used in INGENIAS, which offer less assistance in this process.

4. EXPERIMENTATION

This section illustrates the application of the previous framework (see Section 3). It includes a case study (see Section 4.1) and a report on additional experiments (see Section 4.2).

4.1. Case Study: A Control System for Traffic Lights

The availability of affordable sensors and actuators for ITSs has made traffic experts reconsider some of the traditional designs of traffic facilities. In particular, traffic signals can be made more*intelligent*, in the sense of being aware of the state of traffic in their areas and coordinate with other signals to improve the overall state in wider areas. The use of *intelligent* traffic lights to control junctions in streets is part of this approach [16]. In that work, the *intelligence* comes from two main features of these signals. First, they are able to perceive the number of cars in the lanes that join in their junctions. They achieve this using Optical Information System (OIS) sensors. Second, they exchange information to adapt their behavior to the overall traffic state in the area with the goal of maximizing the throughput of cars from it. In order to evaluate the benefits of the use of these intelligent traffic lights over traditional ones, the work in [16] simulates both situations using the SUMO platform and real data of traffic. This case study reproduces this experiment using the ITS simulation framework.

The process starts with the identification of potential *places* in the problem (see decision 1 and activity 2). According to its definition (see Section 3.1), a *place* is the component that contains *spots*, which are the elements where *containers* can be attached or that *sensors* can monitor and *actuators*act upon. The presentation of the problem introduces several of these *places*. *Junctions* contain intelligent traffic lights where OIS sensors are located. These sensors perceive lanes from *streets*.

At this point, the process can continue by specifying those *places* (starting in decision 3) or the*spots* and related *devices* (starting in decision 14) (see Figure 7). As the original work pays particular attention to the system *devices*, this case study continues in the second path. Figure 8 shows the model instance resulting of it.

Decision 14 begins identifying the *spots* of the previous *places*. The traffic lights are considered as located in the *junction posts* surveying the main *street sections*. Both *junction posts* and *street sections*are *spots*, as the former represent where *devices* are located, and the latter what they survey and affect. Activity 15 specifies these *spots* as *components*. In this case, they do not have any specific state or functionality to specify. Only *street sections* could have it related to the vehicles in them. However, this specification is considered later in the environment *map*.

The next nodes in the path focus on the *devices* the simulated ITS uses to perceive and act. The first step in decision 16 is discovering the *spots* with *containers* for these *devices*. In this case, there is only one of such *spots*: the *junction posts*.

Activity 17 identifies the *devices* running in the *containers*. The *intelligent traffic lights* are modeled as *containers* in *junction posts*. They have as *sensors* the *OIS sensors* that allow them perceiving the queues in their lanes (using the *street sections*); their actuators are their red, green, and yellow*lights*. Activity 18 links *OIS sensors* to the *street sections spots* they watch, and *lights* to the *street sections spots* they are pointing at.

The *containers* of ITSs are not isolated. They set up a complete system thanks to its communications through *channels*. Activity 19 identifies these channels. The *intelligent traffic lights*are connected in this way (not shown in the figure) to share the state of the whole area.

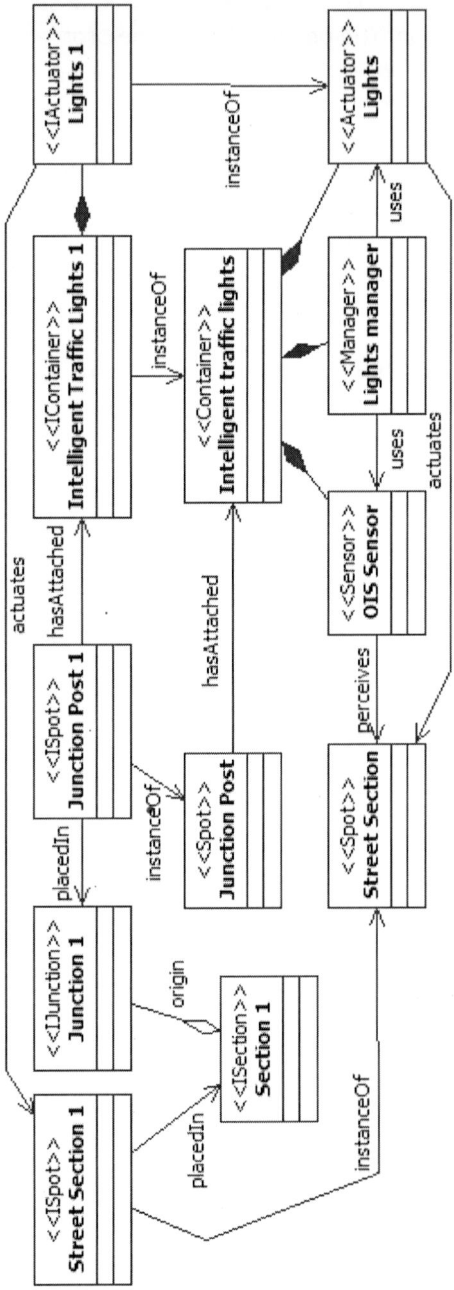

Figure 8. Intelligent traffic lights. *Container* related elements.

Activity 20 models the *sensors* and *actuators* of *containers* as *components*, and their *managers*. An*OIS sensor* has a method to consult the number of vehicles in its lane. A *light* has a state characterized by its current color, the time remaining in it, and the current default time assigned to each color; its methods allow checking the previous values and changing the periods assigned to each color and which to display. As in the SNML, complex control algorithms are modeled using the *managers* of *devices*, which are agents. In [16], traffic lights adapt the time assigned to each color according to the information from their sensors and from traffic lights in the other junctions. For this problem, a *lights manager* orchestrates work in each *intelligent traffic lights* container. It obtains information from its *OIS sensors* and sends it to the *managers* in the other junctions using the*channels*. It also receives from the *channels* the updates from the other *junction posts*. Finally, with all of this information, it updates the lights state.

Figure 8 also shows how to introduce instances of the types identified for a given problem. The instance *lights 1* of the class *lights* runs in the instance *intelligent traffic lights 1* of the *container*, and it governs the traffic of the instance *street section 1* of the class *street section*.

After modeling *spots*, *containers*, and their *devices*, the specification needs to consider the persons and their vehicles, as well as the environment where all these elements interact. This corresponds in the process to the paths for *places* starting in decision 3 (see Figure 7).

In [16], only vehicles are modeled, as drivers are implicit in their control. Using data from real places, researchers determined the volume of traffic in the different sections of streets and the junction turning percentages. In the ITSML, the *vehicle* is a passive *component*, and the decision-making is modeled with the driver *person*.

Activity 5 models the *vehicle component*. In this case, it has no state. The *environment map* models its position (see later activity 13). Its methods should move the car, change the lane, and check its own position, that of other vehicles around it, and the color of the traffic lights in its *street section*.

The specification of *person* is more complex. It includes the elements representing its basic knowledge and actions (see activities 6–9), and the perceive–reflect–act cycle (see activities 10–12).

Activity 6 specifies the *person* as a *component*, mainly regarding the external methods intended to check its state. Here, no one is needed. Activities 7–9 model its knowledge and capabilities. The*goal* of the driver is the path to follow in the street. It is characterized by the entry and exit points to the street, and the exit to take at each junction. The agent only needs to know the position of the vehicle to make decisions. This is a *fact* that directly translates the information provided by the*vehicle* methods. The *tasks* of the agent are those to check the environment and manoeuver the*vehicle*. In the first group are *checking junction* (to know when to turn and in which direction),*checking position* (to know if there is a *vehicle* in a given position), and *checking lights* (to know if the*vehicle* can go into the junction given the traffic lights). The second group includes *moving forward*,*turn* (left or right and the number of exit), and *stop* the *vehicle*.

Activities 10–12 use the previous elements to represent the acting cycle of the driver *person*. Activity 10 models *sense*. It only needs to *perceive* the *vehicle spot* to assert the *facts* regarding its current position, whether the surrounding positions are busy, and the color of traffic lights. Activity 11 models *reflection*. It compares the current position with the next turn. If the *vehicle* has not reached the turning position and lights allow it, it chooses the *task* to move forward; if it has reached the position, it chooses to turn when the target position is available; in other case, it waits. Activity 12 models *acting*, that translates one of the potential *tasks* for execution to *person* and*vehicle methods*.

The last *places* to describe are *things* in the environment. Activity 4 specifies them as *components*. The already mentioned *junctions* and *streets* are *things*. Here, their specifications are empty because these *places* are only containers for *spots*.

The final step in this modeling is activity 13, where *places* are located in the *environment map*. In [16], there are three junctions across a street, and each junction connects two sections of the street and a transversal section from other streets. The *thing street* is located over the three junctions and two sections. These elements and the transversal sections have direct correspondences in the ITSML as parts of the *map*. The model also needs to specify the number and orientation of *lanes* in each *section*.

Figure 9 shows the model resulting of these activities, omitting *lanes* of *sections* and the links between the *street* and *junctions*. All the elements appear as instances (*i.e.*, stereotypes *Iname*) of classes (*i.e.*, stereotypes *name*) that are the direct representation of their meta-classes (*i.e., name* of the meta-class). For instance, the metamodel (see Section 3.1) includes a *section* meta-class. *Section 1–2* and *section transversal 1–3* are instances of a class *section* that is a direct instance (without any modification) of the meta-class *section*. The reason of this modeling is that specifying these instances only needs using existing elements, and not adding new state, methods, or relationships.

When all the previous modeling has finished, activity 21 runs the transformations to create INGENIAS-ML models from these ITSML specifications. Their correspondences follow the inheritance relationships of meta-classes in the ITSML to those of the INGENIAS-ML when they exist. For instance, the ITSML *components* in models become INGENIAS-ML *environment applications*, as *component* is a sub-meta-class of *environment application* (see Figure 3). In the case of multiple inheritances, the base target meta-class is that with the closer semantics. For instance, the ITSML *person* inherits from the INGENIAS-ML *agent*, and through *place* and *component* also of the INGENIAS-ML *environment application* (see Figure 3 and Figure 4). However, its meaning is closer to the INGENIAS-ML *agent*. Thus, the transformation generates for a *person* an INGENIAS-ML*agent*, though it also adds additional elements corresponding to it being also a *place*.

Finally, activity 22 follows an INGENIAS process [27,28] to refine those INGENIAS models and generate the source code for the simulation in SUMO [16]. In this case, INGENIAS models mainly add the description of the interactions among *manager* agents regarding the exchange of information from their *intelligent traffic lights*. SUMO specifies most of the components of its simulations using XML files. The main mappings with the original concepts from the ITSML are the following. The *map* of the *environment* is represented in SUMO as a *network*. Its *nodes* and *edges*are, respectively, ITSML *junctions* and *lanes*. The SUMO *vehicle* merges the ITSML *vehicle* and driver*person*. SUMO does not allow controlling the movements of its *vehicles*, as it provides only a limited set of implementations for them. Nevertheless, the path vehicles have to follow can be specified, and SUMO controls that traffic signals are observed. Regarding traffic lights, SUMO supports*actuated traffic lights* for the experiments described in [16]. They can change the time showing each color according to the number of vehicles in the *edges* they control.

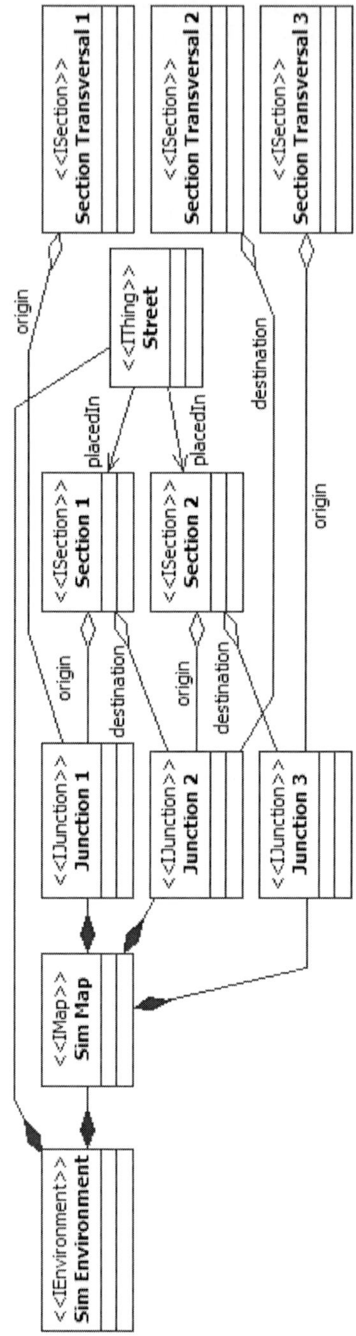

Figure 9. Intelligent traffic lights. *Environment* related elements.

4.2. Discussion

The proposed framework has been tested in several experiments with the purpose of assessing its components. These experiments considered different contexts in order to evaluate its adaptability to multiple needs regarding simulation. In particular, the ITSML has been used to specify different models for people and types of ITS, and the process and tools also for several target platforms.

Regarding decision-making (see Section 3.1.3), there has been work with the goal-model of INGENIAS [14] and the reactive model for drivers in [32]. Compared to the ITSML, INGENIAS offers additional primitives to model goals with the distinction between hard (with quantitative evaluations) and soft (with qualitative evaluations) goals, and different relationships (e.g., affects, contributes positively, satisfies, and AND and OR compositions). The work in [32] offers an alternative perceive–reflect–act cycle. Its agents consider timed data, and combine for reflection a reactive model and a short-time planner.

The concepts in the previous models are specializations of types already available in the ITSML. For instance, hard and soft goals [14] are sub-types of *goal*, and the short-time planner [32] of*reflection*. Relationships can be represented using instances of the *relatedTo* meta-relationship. Nevertheless, this is a limited approach, as it does not allow introducing constraints. For instance, INGENIAS relations between goals can be modeled, but not constrained with graphical primitives to link only goals.

There is also work with other types of ITS. Smart roads [33] are ITS which SNs are mainly deployed on roads. Their models with the ITSML use specializations of concepts related to devices and managers (see Section 3.1.2) and the environment (see Section 3.1.4). In [33], there are eye cats (a type of *spot*) deployed on road borders (which are *things*). The eye cats hold magnetic trackers (as *sensors*) that perceive vehicles passing near them. *Manager agents* represent the control of these sensors and their communications.

The framework has also been tested with different target simulation platforms. Besides the already mentioned SUMO [16], it has also been tested with MATSim [34] and the platform for traffic simulation in [35]. MATSim mainly defines vehicle movement using paths and additional parameters that influence it, as SUMO does. Thus, the approach described for SUMO in the case study (see Section 4.1) is applicable with minor changes to MATSim. MATSim can also be extended with additional behavioral modules, as the platform in [35]. The latter distinguishes between vehicles and drivers, and supports complex agent-based models for people. Drivers have a basic behavior of path following that can incorporate maneuvers and general constraints on movement as goals. With the ITSML, those constraints are also modeled as *goals*, linked to *tasks* that represent maneuvers. Tasks are related to *conditions* that represent when those maneuvers can be considered for execution. In all the cases, the first tests with these platforms require developing specific code snippets to connect already available code (that for abstract models) with the platform. This also happened in the experiment reported in the case study. These new fragments could later be reused, or adapted with less effort, when experiments continued, and integrated in the code generation tasks.

In these cases, the process points out the ITSML primitives suitable to model the concepts from those models and platforms. However, two aspects remain little described, the choice of relationships and the use of inheritance. Relationships are deduced from the context provided by concepts. The process focuses on the latter, and these determine the allowed relationships. There are no indications on how to choose when several relationships are available beyond their description in the ITSML. There are also sometimes difficulties to design inheritance hierarchies, particularly to establish what concepts must be extended.

5. QUALITATIVE COMPARATIVE ANALYSIS

This work is related to two main aspects of research in ITSs. It deals with the relevant concepts to model this kind of system, and the approaches followed in their analysis, development, and simulation.

ITSs cover a wide variety of needs [3,5,6,36], e.g., reduction of traffic congestions, toll collection, parking, collision avoidance, or emergency response. However, their architectures and perspectives on the environment present a similar structure. There is a world external to the system that includes, for instance, environmental conditions (e.g., weather, sun position, or daytime) [33,37], roads [33,38], vehicles [37,39], or pedestrians [39]. ITSs include devices (sensors and actuators), which are their “interface” to interact with that environment [6,40]. Using them, ITSs provide their services. These services may require computational resources beyond devices, e.g., servers to store information or allow users to monitor the system and the environment.

The previous facets can be effectively integrated using a view of these elements based on external interfaces able to manipulate their state. In most cases, their inner structure and behavior can be disregarded. For instance, with the exception of works focused on device control (e.g., [39,41]), most of research adopts a high-level view of devices, focused on the data they gather and the services they provide (e.g., [33,38,40]). Such perspective is also applied in the models of environment elements when needed. The only exception is the map of the environment, which is modeled with some kind of graph, e.g., when working with location information from the Global Positioning System (GPS) [38].

The point of view of simulation is slightly different. It needs to integrate concepts to model the objects in the external reality and their behavior, so it can represent their influence in sensors and actuators [42]. For instance, drivers' behavior is not only an aspect to observe, but also to manipulate in order to test the person and system functioning in multiple circumstances.

Regarding vehicles, simulations need at least to establish partially their path (*i.e.*, origin and destination, and sometimes intermediate points). The specific way in which a vehicle meets this objective offers variations. There are works that consider an strict adherence to the path (*i.e.*, the path constitutes the precise movement instructions) [16], some partially random movements around it (e.g., to decide how to move between two points of the target path) [43], path calculation algorithms (e.g., shortest path under different cost functions [16,34,43]), and specific maneuvers (e.g., car following, lane change, or turns) [43,44] to follow it. Nevertheless, microscopic models (where vehicles are individually modeled) such as [16,34,43,44], could allow modeling more complex and flexible behaviors by replacing the default ones. However, this is not documented in the reviewed works, and it would probably require modifying the code of the considered simulation platforms.

Most of works discard the explicit modeling of drivers, and only consider vehicle control [16,43]. Some references [43,44] include pedestrians, but only as moving objects, similar in their description to vehicles. General traffic studies are more prone to consider these issues. Their models can specify the mental state and features of people, for instance regarding their goals and ability [30], their decision-making [45], or their roles in traffic (*i.e.*, pedestrians, drivers, and passengers) [12].

Both in the case of vehicles and people, researchers lack of support to extend the provided default implementations. For instance, there is little documentation, libraries, base classes for inheritance, or schedulers for multiple tasks, in the platforms to help coding these modifications.

Our ITSML addresses these issues at several levels. It provides an extensible metamodel able to accommodate the different previous modeling approaches. The specification of environment elements and vehicles as objects with state and method interfaces is achieved with *components*. People and vehicle movements are modeled through the specification of *persons*. The case study (see Section 4.1) already includes an example of path following with maneuvers. The adoption of*agents* to represent *persons* goes further and allows describing the wide range of alternative models for people in traffic studies, from reactive to deliberative ones. Moreover, *managers* of devices are also modeled as *agents*, which facilitates representing complex distributed algorithms for ITSs including information management (e.g., goals, facts, probabilities, or uncertainty). Many issues usually considered in code are promoted to models, so researchers can define their requirements using MLs. Then, standard MDE development methodologies translate models to source code. When this translation is done, it provides documentation and reusable artifacts (*i.e.*, models and transformations) for similar implementations. Thus, it provides the basis to build improved support resources for these tasks.

The already discussed conceptual frameworks are intended to be the basis for the analysis of ITSs. They focus on the external interface of these systems (*i.e.*, behavior perceived by users), the environment, and their mutual influences. However, those references do not consider the internal organization of such systems, beyond, for instance, coding constraints imposed by simulation platforms. There are specific works focused on the architecture of ITSs. They usually present abstract models and hints for their implementation (e.g., [33,37]). Other works address low-level aspects of these systems, e.g., control or coordination mechanisms. They are usually focused on specific aspects, e.g., issues of computer vision in [41]. When dealing with such details, the use of simulations is common [7,16,42].

Works such as SUMO [16], DEUS [43], MATSim [34], the Tokyo Virtual Living Lab [44], the test platform in [35], or those reviewed in [42], are traffic simulation platforms. They include libraries to develop and run simulations, and sometimes user interfaces to specify them. Although there are descriptions of their underlying abstract theoretical models, they are black-boxes. A complete understanding of their functioning can require examining their source code when it is available. Literature has already pointed out this issue as one of the main limitations in the use of simulations for research, as it makes difficult guaranteeing the alignment between abstract models and simulations [8].

In most cases, works do not describe the development process followed. There are not explicit guidelines to reproduce modeling, design, or coding. This lack of documentation hampers validation and sharing experiences on development. Moreover, works focus on solving specific issues at a given level of abstraction, e.g., abstract model, architecture, or control of devices. They do not identify ways to integrate that research with complementary works to offer a complete solution to develop ITSs.

The current work adopts MDE to support these tasks. It provides an approach with several abstraction levels intended for different backgrounds. Researchers on ITSs work with the ITSML at the levels of CIM and PIM. Designers of simulations and programmers work mainly with the INGENIAS-ML at the levels of PIM and PSM. In both cases, experts integrate and refine models adding new information and using transformations. For the first group of tasks, this work offers modeling guidelines; for the second, INGENIAS provides a complete development methodology. The overall approach facilitates the explicit representation of all the information involved in the transition from the abstract models to code, which facilitates validation and reutilization.

6. CONCLUSIONS

This paper has presented a MDE approach for the analysis through simulations of ITSs. The complexity of these systems and the environments where they act implies high development costs. This kind of issues has been addressed in other domains through simulation. However, its application demands means to facilitate the transition from the abstract models of systems to the simulation, and its validation. Usually, researchers make these tasks manually, what is an additional source of mistakes. Adopting MDE approaches, researchers explicitly describe these steps with models and transformations. This facilitates their examination, and therefore validation and testing.

The proposed framework comprehends a ML, guidelines, and tools to develop ITS specifications and simulations. It is intended for the complete development process, focusing effort on the high-level design. Its definition reuses previous research in aspects related to ITSs. The ITSML is built over the TML [12] and the SNML [13], extending them with ITS-specific concepts. All these languages adopt the agent paradigm [11], which facilitates integrating the different perspectives.

The ITSML conceives all the elements in this domain as *components*, which are characterized by their state and a method-based interface. Flows of control and information among components use these methods. ITSs interact with their external environment using a layer of sensors and actuators, which are also *components*. Complex algorithms are modeled with *agents*. These include components for perception, reflection, and acting that manipulate information and goals.

For control activities, *sensors* perceive *events* from the environment, which they transform in internal *facts*. Their *methods* generate *notifications* from those *facts* that are communicated to other *components*, including *manager* agents. These use available information to check *goals* and decide what *tasks* to attempt. The execution of *tasks* invokes *methods*, which can generate new *notifications*. Some of these *notifications* trigger the execution of *actuator methods*, which act on the environment.

The concepts related to traffic are organized following an adapted DVE [30] model according to the structure of the TML [12]. They include common primitives such as *person*, *vehicle*, and *thing* (*i.e.*, environment element). Here, they are linked to the group of concepts corresponding to *devices* and their *manager* controllers extracted from the SNML. This last group models the interface between ITSs and their external environment.

The ITSML also provides mechanisms of inheritance and instantiation of classes in models. The first one allows flexible extensions of the ML without modifying its metamodel. The second one supports the complete definition of the components of the simulation at the model level. In this way, only algorithms need to be specified with code.

The development guidelines, along with the definition of concepts in the metamodel, help designers to build their specifications. They provide a step-by-step process to look for relevant elements in problems, and translate them to modeling primitives. The low-level design and coding follow INGENIAS [14], an existing MDE methodology for MASs. The use of agent concepts in the ITSML and the INGENIAS-ML facilitates the transition between both stages using automated transformations.

The final elements of the framework are the support tools. A model editor based on INGENME [15] supports specifying models compliant with the ITS metamodel. The code generator of the traffic framework [12] is used in the later stages of the INGENIAS process. It replaces the usual INGENIAS tools, because it provides more guidance to work jointly with models and code templates.

This framework has been used in several experiments. The case study developed a simulation of an ITS with intelligent traffic lights for the simulation platform SUMO [16]. The models specified most of the intended simulation. Only algorithms (e.g., using thresholds and timing to change the color of traffic lights, or the methods to interact with SUMO) need to be coded. As most of this code appears in templates that can be reused in code generation for different model elements, the actual coding effort is highly reduced. Changes in the simulation would only affect the model if they add new instances of available classes of entity or relationship. If new classes appear, only the templates for them need to be added before regenerating the simulation. Templates could be reused from other simulation projects. Similar findings appear in the other reported experiments, which considered simulations with extended models of decision-making, other types of ITS, and several simulation platforms.

This framework is still ongoing work with several open issues, some of them already pointed out in experiments in Section 4. First, the metamodel needs to be enriched with additional primitives. Primitives to represent *tasks* ordering or preferences are not available. Also, the specification of *vehicle* parts and their actions is not possible, for instance to represent the angle and acceleration of a vehicle turn in a junction as done in [44]. Second, certain constraints on the model, for example in the number of instances of a class, inheritance relationships, or values of attributes, cannot be expressed. The use of a constraint language such as the Object Constraint Language (OCL) [46] would reduce here the need to resort to code. Third, the development guidelines need to be extended. Now, they indicate a possible ordering of steps and leave for the metamodel definition the indications on how to identify the different modeling concepts. The identification hints should also be transferred to the process specification. Finally, additional experimentation is required to validate the framework.

Current experiments are limited in number, the types of ITS modeled, and the considered simulation infrastructures.

ACKNOWLEDGMENTS

This work has been done in the context of the projects “Social Ambient Assisting Living—Methods (SociAAL)” (TIN2011-28335-C02-01) supported by the Spanish Ministry for Economy and Competitiveness. Also, we acknowledge support from the “Red Científico-Tecnológica en Ciencias de los Servicios” (TIN2011-15497-E) and the “Programa de Creación y Consolidación de Grupos de Investigación” (UCM-BSCH GR35/10-A).

AUTHOR CONTRIBUTIONS

Alberto Fernández-Isabel developed the framework for the MDE of ITS simulations, and implemented the case study. Rubén Fuentes-Fernández adapted the INGENIAS and INGENME infrastructure for the framework. All the authors contributed equally to writing the paper.

REFERENCES

1. Chourabi, H.; Nam, T.; Walker, S.; Gil-Garcia, J.R.; Mellouli, S.; Nahon, K.; Pardo, T.A.; Scholl, H.J. Understanding Smart Cities: An Integrative Framework. In Proceedings of the 45th Annual Hawaii International Conference on System Sciences (HICSS 2012), Maui, HI, USA, 4–7 January 2012; Sprague, R.H., Jr., Ed.; IEEE: Washington, DC, USA, 2012; pp. 2289–2297.

2. Harrison, C.; Donnelly, I.A. A Theory of Smart Cities. In Proceedings of the 55th Annual Meeting of the International Society for the Systems Sciences (ISSS 2011), Hull, UK, 17–22 July 2011; pp. 1–15.

3. Tewolde, G.S. Sensor and Network Technology for Intelligent Transportation Systems. In 2012 IEEE International Conference on

Electro/Information Technology (EIT 2012), Indianapolis, IN, USA, 6–8 May 2012; IEEE: Washington, DC, USA, 2012; pp. 1–7.

4. The European Parliament and the Council of the European Union. Directive 2010/40/EU of the European Parliament and of the Council of 7 July 2010, on the Framework for the Deployment of Intelligent Transport Systems in the Field of Road Transport and for Interfaces with Other Modes of Transport. Available online: http://eur-lex.europa.eu/(accessed on 15 March 2015).

5. Figueiredo, L.; Jesus, I.; Machado, J.A.T.; Ferreira, J.; de Carvalho, J.L.M. Towards the Development of Intelligent Transportation Systems. In Proceedings of the 2001 IEEE Intelligent Transportation Systems Conference (ITSC 2001), Oakland, CA, USA, 25–29 August 2001; IEEE: Washington, DC, USA, 2001; pp. 1206–1211.

6. Klein, L.A. *Sensor Technologies and Data Requirements for ITS*; Artech House Publishers: Norwood, MA, USA, 2001.

7. Pursula, M. Simulation of Traffic Systems—An Overview. *J. Geogr. Inf. Decis. Anal.* **1999**, *3*, 1–8.

8. Axtell, R.L.; Epstein, J.M. Agent-based modeling: Understanding our creations. *Bull. St. Inst.***1994**, *9*, 28–32.

9. Bézivin, J. Model Driven Engineering: An Emerging Technical Space. In *Generative and Transformational Techniques in Software Engineering*; Lecture Notes in Computer Science; Lämmel, R., Saraiva, J., Visser, J., Eds.; Springer: Heidelberg, Germany, 2006; Volume 4143, pp. 36–64.

10. Fuentes-Fernández, R.; Hassan, S.; Pavón, J.; Galán, J.M.; López-Paredes, A. Metamodels for role-driven agent-based modelling. *Comput. Math. Organ. Theory* **2012**, *18*, 91–112.

11. Müller, J.P.; Pischel, M.; Thiel, M. Modeling Reactive Behaviour in Vertically Layered Agent Architectures. In *Intelligent Agents*; Lecture Notes in Computer Science; Wooldridge, M.J., Jennings, N.R., Eds.; Springer: Heidelberg, Germany, 1995; Volume 890, pp. 261–276.

12. Fernández-Isabel, A.; Fuentes-Fernández, R. A Model-Driven Engineering Process for Agent-Based Traffic Simulations. In Proceedings of the 5th International Conference on Simulation and Modeling

Methodologies, Technologies and Applications (SIMULTECH 2015), Colmar, France, 21–23 July 2015; SCITEPRESS: Setúbal, Portugal, 2015; pp. 1–10.

13. Fuentes-Fernández, R.; Guijarro, M.; Pajares, G. A Multi-Agent System Architecture for Sensor Networks. *Sensors* **2009**, *9*, 10244–10269.

14. Pavón, J.; Gómez-Sanz, J.J.; Fuentes, R. The INGENIAS Methodology and Tools. In *Agent-Oriented Methodologies*; Henderson-Sellers, B., Giorgini, P., Eds.; Idea Group Publishing: Hershey, PA, USA, 2005; Chapter IX; pp. 236–276.

15. Pavón, J.; Gómez-Sanz, J.; Paredes-López, A. The SiCoSSyS approach to SoS engineering. In Proceedings of the 6th International Conference on System of Systems Engineering (SoSE 2011), Albuquerque, NM, USA, 27–30 June 2011; IEEE: Washington, DC, USA, 2011; pp. 179–184.

16. Krajzewicz, D.; Brockfeld, E.; Mikat, J.; Ringel, J.; Rössel, C.; Tuchscheerer, W.; Wagner, P.; Wösler, R. Simulation of modern Traffic Lights Control Systems using the open source Traffic Simulation SUMO. In Proceedings of the 3rd Industrial Simulation Conference (ISC 2005), Berlin, Germany, 9–11 June 2005; Krueger, J., Lisounkin, A., Schreck, G., Eds.; EUROSIS-ETI: Ostend, Belgium, 2005; pp. 299–302.

17. Object Management Group. *OMG Meta-Object Facility (MOF) Core Specification, Version 2.4.2*; Object Management Group, 2014. Available online: http://www.omg.org/ (accessed on 15 March 2015).

18. Object Management Group. *OMG Unified Modeling Language (OMG UML), Version 2.5*; Object Management Group, 2013. Available online: http://www.omg.org/ (accessed on 15 March 2015).

19. Steinberg, D.; Budinsky, F.; Paternostro, M.; Merks, E. *EMF—Eclipse Modeling Framework*, 2nd ed.; Addison-Wesley Professional: Boston, MA, USA, 2008.

20. Lopes, D.; Hammoudi, S.; Bézivin, J.; Jouault, F. Mapping Specification in MDA: From Theory to Practice. In *Interoperability of Enterprise Software and Applications*; Konstantas, D., Bourrières, J.-P., Léonard, M., Boudjlida, N., Eds.; Springer: Heidelberg, Germany, 2006; pp. 253–264.

21. Kelly, S.; Lyytinen, K.; Rossi, M. Metaedit+: A fully configurable multi-user and multi-tool CASE and CAME environment. In *Advanced Information Systems Engineering*; Lecture Notes in Computer Science; Constantopoulos, P., Mylopoulos, J., Vassiliou, Y., Eds.; Springer: Heidelberg, Germany, 1996; Volume 1080, pp. 1–21.

22. Mens, T.; van Gorp, P. A Taxonomy of Model Transformation. *Electron. Notes Theor. Comput. Sci.* **2006**, *152*, 125–142.

23. Czarnecki, K.; Helsen, S. Feature-Based survey of model transformation approaches. *IBM Syst. J.* **2006**, *45*, 621–645.

24. Eclipse M2T. Eclipse Modeling-M2T-Home. Available online: https://www.eclipse.org/modeling/m2t/ (accessed on 15 March 2015).

25. Eclipse MMT. Eclipse Model to Model Transformation. Available online: https://www.eclipse.org/mmt/ (accessed on 15 March 2015).

26. Kleppe, A.G.; Warmer, J.; Bast, B. *MDA Explained: The Model Driven Architecture—Practice and Promise*; Addison-Wesley: Boston, MA, USA, 2003.

27. Gómez-Rodríguez, A.; Fuentes-Fernández, R.; González-Moreno, J.C.; Rodríguez-Martínez, F.J. INGENIAS with the Unified Development Process. In *Handbook on Agent-Oriented Design Processes*; Cossentino, M., Hilaire, V., Molesini, A., Seidita, V., Eds.; Springer: Heidelberg, Germany, 2014; pp. 371–405.

28. González-Moreno, J.C.; Gómez-Rodríguez, A.; Fuentes-Fernández, R.; Ramos-Valcárcel, D. INGENIAS-Scrum. In *Handbook on Agent-Oriented Design Processes*; Cossentino, M., Hilaire, V., Molesini, A., Seidita, V., Eds.; Springer: Heidelberg, Germany, 2014; pp. 219–251.

29. García-Magariño, I.; Fuentes-Fernández, R. A Technique for Defining Metamodel Translations. *IEICE Trans. Inf. Syst.* **2009**, *92*, 2043–2052.

30. Amditis, A.; Pagle, K.; Joshi, S.; Bekiaris, E. Driver-Vehicle-Environment monitoring for on-board driver support systems: Lessons learned from design and implementation. *Appl. Ergon.* **2010**, *41*, 225–235.

31. Rubel, D.; Wren, J.; Clayberg, E. *The Eclipse Graphical Editing Framework (GEF)*; Addison-Wesley Professional: Boston, MA, USA, 2011.

32. Ehlert, P.A.M.; Rothkrantz, L.J.M. Microscopic traffic simulation with reactive driving agents. In Proceedings of the 2001 IEEE Intelligent Transportation Systems Conference (ITSC 2001), Oakland, CA, USA, 25–29 August 2001; IEEE: Washington, DC, USA, 2001; pp. 860–865.

33. Karpiriski, M.; Senart, A.; Cahill, V. Sensor networks for smart roads. In Proceedings of the 4th Annual IEEE International Conference on Pervasive Computing and Communications, Workshops (PerComW 2006), Pissa, Italy, 13–17 March 2006; IEEE: Washington, DC, USA, 2006; p. 310.

34. Transport Systems Planning and Transport Telematics Group (Technische Universität Berlin); Transport Planning Group (Swiss Federal Institute of Technology Zurich); Senozon Company. *MATSim, Multi-Agent Transport Simulation*; MATSim, 2015. Available online: http://www.matsim.org/ (accessed on 25 May 2015).

35. Fernández-Isabel, A.; Fuentes-Fernández, R. An Agent-Based Platform for Traffic Simulation. In Soft Computing Models in Industrial and Environmental Applications, 6th International Conference SOCO 2011; Advances in Intelligent and Soft Computing. Corchado, E., Snášel, V., Sedano, J., Hassanien, A.E., Calvo, J.L., Ślęzak, D., Eds.; Springer: Heidelberg, Germany, 2011; Volume 87, pp. 505–514.

36. Pascale, A.; Nicoli, M.; Deflorio, F.; Dalla Chiara, B.; Spagnolini, U. Wireless sensor networks for traffic management and road safety. *IET Intell. Transp. Syst.* **2012**, *6*, 67–77.

37. Sun, Z.; Bebis, G.; Miller, R. On-Road vehicle detection: A review. *IEEE Trans. Pattern Anal. Mach. Intell.* **2006**, *28*, 694–711.

38. Faouzi, N.-E.E.; Leung, H.; Kurian, A. Data fusion in intelligent transportation systems: Progress and challenges—A survey. *Inf. Fusion* **2011**, *12*, 4–10.

39. Gandhi, T.; Trivedi, M.M. Pedestrian protection systems: Issues, survey, and challenges.*IEEE Trans. Intell. Transp. Syst.* **2007**, *8*, 413–430.

40. Hancke, G.P.; de Carvalho e Silva, B.; Hancke, G.P., Jr. The Role of Advanced Sensing in Smart Cities. *Sensors* **2012**, *13*, 393–425.

41. Trivedi, M.M.; Gandhi, T.; McCall, J. Looking-In and looking-out of a vehicle: Computer-Vision-Based enhanced vehicle safety. *IEEE Trans. Intell. Transp. Syst.* **2007**, *8*, 108–120.

42. Kotusevski, G.; Hawick, K.A. A Review of Traffic Simulation Software. *Res. Lett. Inf. Math. Sci.* **2009**, *13*, 35–54.

43. Picone, M.; Amoretti, M.; Zanichelli, F. Simulating Smart Cities with DEUS. In Proceedings of the 5th International ICST Conference on Simulation Tools and Techniques (SIMUTOOLS 2012), Desenzano, Italy, 19–23 March 2012; pp. 172–177.

44. Prendinger, H.; Gajananan, K.; Zaky, A.; Fares, A.; Molenaar, R.; Urbano, D.; van Lint, H.; Gomaa, W. Tokyo Virtual Living Lab: Designing Smart Cities Based on the 3D Internet. *IEEE Internet Comput.* **2013**, *17*, 30–38.

45. Salvucci, D.D.; Boer, E.R.; Liu, A. Toward an Integrated Model of Driver Behavior in Cognitive Architecture. *Transp. Res. Rec. J. Transp. Res. Board* **2001**, *1779*, 9–16.

46. Object Management Group. *Object Constraint Language, Version 2.4*; Object Management Group, 2014. Available online: http://www.omg.org/ (accessed on 15 March 2015).

Chapter 4

Architecture of a Framework for Providing Information Services for Public Transport

Carmelo R. García *, Ricardo Pérez, Álvaro Lorenzo, Alexis Quesada-Arencibia, Francisco Alayón and Gabino Padrón

Institute for Cybernetics, University of Las Palmas de Gran Canaria, Campus de Tafira, Las Palmas de Gran Canaria, 35017 Las Palmas, Spain

ABSTRACT

This paper presents OnRoute, a framework for developing and running ubiquitous software that provides information services to passengers of public transportation, including payment systems and on-route guidance services. To achieve a high level of interoperability, accessibility and context awareness, OnRoute uses the ubiquitous computing paradigm. To guarantee the quality of the software produced, the reliable software principles used in critical contexts, such as automotive systems, are also considered by the framework. The main components of its architecture (run-time, system services, software components and development discipline) and how they are deployed in the transportation network (stations and vehicles) are described in this paper. Finally, to illustrate the use of OnRoute, the development of a guidance service for travellers is explained.

KEYWORDS

ubiquitous computing frameworks; service-orientation; intelligent transport systems

1. INTRODUCTION

This paper falls within the context of intelligent public transport systems, particularly the technologies for developing and running software systems that provide information services to users of public transportation. One way to improve transport systems is to provide passengers with useful information services. Often, these services require complex software systems that need a high degree of interoperability because they have to operate in a wide range of technological contexts and serve a massive number of users, as is the case with large public transport systems. Therefore, it is advisable to have platforms and environments that facilitate the production and execution of reliable software for these services.

In this paper, we describe a framework, known as OnRoute, for producing and running software that provides information services for public transport passengers. The environment provides a set of operating principles and components from which services, such as on-route guided assistance, information at stops and payment systems, can be developed. For the purposes of OnRoute, data relating to the position of the vehicle and planning, as well as the local communications system, are the components demanded of the on-board infrastructure. OnRoute services are accessible through users' mobile devices, e.g., mobile telephones and Smart Phones, using Bluetooth communication infrastructure that is installed at important points of the transport network, such as stations, stops and vehicles. This framework has been used to develop three examples of information services. The first example is an on-route guidance assistant that provides the passengers of a public transport bus with information pertaining to points of interest along the route, the next stop and the estimated time of arrival at the next stop or the final stop. The second example is a payment system that allows travellers to pay the bus fare. The third example is an inspection system that permits the transport company staff to access vehicle information on route to validate the number of passengers on-board and the different operations conducted during travel. In all cases, users have to use their mobile devices (mobile or smart phones) to access the services provided by the public transport vehicle.

OnRoute uses the paradigms of ubiquitous computing [1] to produce and provide support to software that offers public transport information services aimed at users who interact in contexts of mobility, more specifically, passengers, drivers and operations control staff. Services of this type require the resources used by the transport network infrastructure to automate their production activities. We first highlight the on-board systems that encompass all of the devices installed on board the buses to control the services that the vehicle offers, support the payment systems and handle the relevant events that occur during the operations conducted in the vehicle. For the purposes of OnRoute, data relating to the position of the vehicle and planning along with the local communications system are the components demanded of the on-board infrastructure. OnRoute also uses the local communications infrastructure installed at important points of the transport network, such as stations and stops. Specifically, the communication infrastructures used by OnRoute are Bluetooth for communication with user mobile terminals and Wifi for communication with the transport infrastructure. For example, a vehicle must always adhere to the timetable, reaching each stop at the correct time. If the vehicle does not, the on-board system must detect this planning breach and report it in real time to take the necessary measures to minimise the negative impact on the customer. Such measures would include communication with the OnRoute information services using the vehicle's Wifi infrastructure.

This paper is structured as follows: Section 2 describes a set of works related to the topics of this paper; the technology used is presented in the third section. The fourth section is dedicated to explaining the requirements of the OnRoute framework. The OnRoute architecture is described in Section 5, which explains the main elements of the architecture from two points of view: operation principles and software development. Section 6 offers an illustration of how to use the OnRoute framework to implement an on-route guidance assistant. Finally, the main conclusions and future work are presented in the last section.

2. RELATED WORKS

The potential offered by intelligent transport systems for improving public transport is described in many of the papers listed in the bibliography. For example, Giannopoulus [2] analyses how information and communication technology can be used to develop systems that offer passengers information, adapting it to different operating environments and to different kinds of passengers.

The development of ubiquitous or pervasive systems in the context of transport falls into three categories: contributions that describe frameworks, works that propose an alternative model based on middleware, and works that propose models for the development of pervasive applications that require codes to implement planning and optimisation algorithms. In the first group, Meier [3] proposes the iTransIT framework to integrate transport systems that model spatial information and to implement the spatial application programming interface for pervasive information services, such as smart traveller information services. In the case of middleware, Gluli [4] describes a service-oriented middleware solution and its implementation, in a demonstration vehicle, assuming the non-functional requirements of security, privacy, usability, and reliability (SPUR). In the third group, Ossowski *et al.* [5] introduce a social multi-agent decision support system architecture and provide design guidelines to construct an agent-based decision support system. To illustrate the method, it is applied to two classic transport problems: road traffic management and bus fleet management. Harrington and Cahill [6] propose a model-driven approach to developing pervasive applications. In particular, these authors address how to generate code to implement planning and optimisation algorithms in pervasive environments. These authors present an empirical evaluation of the model impact on the development of a pervasive application in the domain of intelligent transportation systems, such as the optimisation of traffic light settings in an urban traffic control system to minimise the waiting time for vehicles.

In the context of ubiquitous software, Lopez de Ipiña *et al.* [7] have devised a framework for the development and deployment of smart objects and the transformation of a user's mobile device into a universal remote controller. Barretto [8] proposes a framework for developing and running sensor-based, context-aware agents that are based on re-usable components, while Römer [9] proposes a framework based on Jini and Web Services. García-Herranz [10] proposes a programming system that allows the end users to control and programme their environments through a uniform, application–independent method.

3. DOMAIN MODEL OF ONROUTE: SYSTEM REQUIREMENTS

The system that we describe in this paper is a framework to programme and execute mobile applications that provide information services in the public transport context. These information services have a high degree of accessibility by the users (travellers, transport company staff and regulatory agency staff). The services provided by applications that have been developed using OnRoute are accessible through general-purpose mobile terminals that do not necessarily perform well. Therefore, access to the services must be through local wireless communication networks that are available to these kinds of devices. Moreover, the mobile device applications developed using OnRoute are Java applications built using the Java Platform Micro Editions (the applications developed by the OnRoute framework will run in the public transport domain), the proposed architecture meets the following requirements:

- Heterogeneity of mobile devices. Services should be available to a variety of mobile devices.

- Scalability. The system allows new elements to be added to the infrastructure that permit newly developed information services to be added or make them accessible to a greater number of users.

- Spontaneous interaction. The system allows for the spontaneous interaction with users that are already using other system services; this number of users is potentially massive.

These OnRoute requirements can be structured into two groups. The first group is related to the general requirements of the ubiquitous systems, and the second is related to the requirements of the transport information systems.

In terms of the requirements of the ubiquitous systems, the software produced using OnRoute can be characterised by its capacity to integrate its surrounding physical and technological environment. Consequently, it can operate autonomously and spontaneously in different environments.

To attain these functionalities, OnRoute accepts the principles that characterise pervasive system software [11]. The boundary principle establishes that the distinction between environments in pervasive frameworks must be made by boundaries that mark differences in content, and these boundaries do not need to limit the interoperability of the systems. The principle of volatility establishes that pervasive systems must accept that the number of users, devices and applications that intervene in a pervasive environment is unpredictable. For this reason, a set of invariable operating principles that govern the running of the system must be specified. Because of these characteristics, the OnRoute architecture is deployed in two areas. The first area is the infrastructure that the public transport system provides. This includes a basic set of components, comprising all of the elements that allow user applications to access transport-related information. The second area is user devices, comprising all of the components that have the capacity to integrate into the different environments and that facilitate access to the information produced by the OnRoute information services.

Following Hervas [12], ontology provides the following benefits: it enables the system to be interoperable, it reduces the difficulties related to technological diversity, and it facilitates communication between humans and computing systems. To facilitate the identification of specific requirements related to the transport information system, the main ontological elements of OnRoute are presented below:

- Line. This is the route followed by the vehicles. During the route, the information services are provided to the passengers.

- Vehicle. This is the element used to transport passengers. When a vehicle is in service, it follows a route. During the route, information services are provided to the passengers.

- Service. This is an abstract entity that models transport data. It offers useful information services to the passengers, such as a payment system or an on-route guidance assistant.

- Programmer. This is the person who develops a service and defines the data and the functionalities contained in the service.

- Mobile terminal. A device used by the transport client to access the information services.

- OnRoute. The active entity, executed in vehicles, that provides information services.

- Bluetooth interface. The interface used by mobile terminals to communicate with OnRoute and *vice versa*.

The relationships between these main concepts of the OnRoute framework are shown in Figure 1.

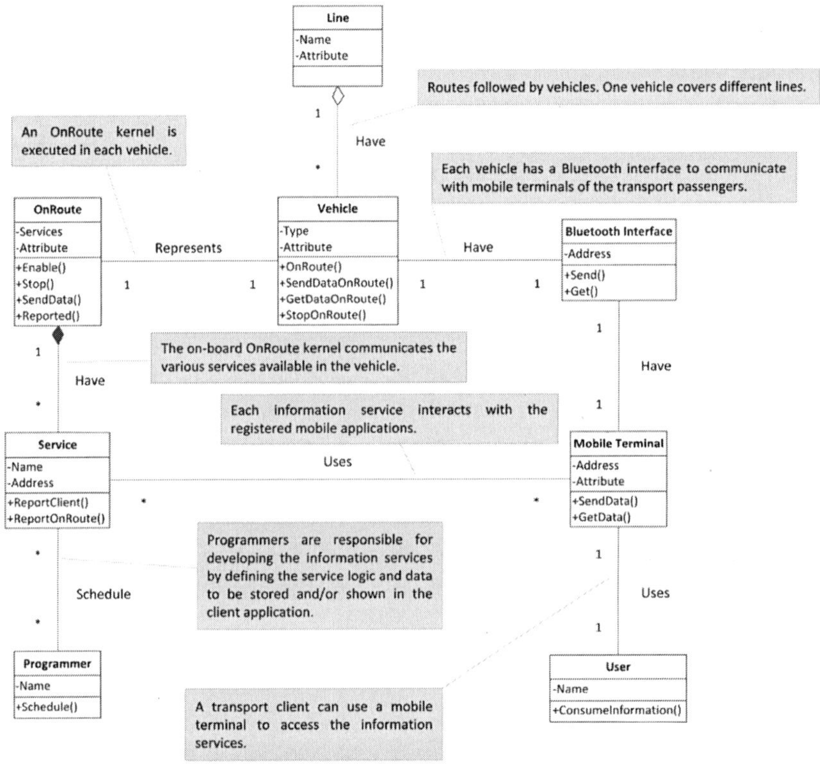

Figure 1. Relations between main ontological concepts.

4. ONROUTE ARCHITECTURE

OnRoute provides a framework for developing and running software that offers information services to public transport passengers using the existing infrastructures available on board vehicles at stops and in stations.

Passengers use their computing and communications devices (mobile telephones and Smart Phones) to access the information generated using OnRoute (seeFigure 2). We will have a set of OnRoute services with varying availability and a set of OnRoute client applications that will vary in number. If a passenger is in a station that is busy in terms of both passengers and vehicles, the number can be massive. This situation poses the greatest difficulty for meeting all of the operating requirements and principles described in this section. The reason for this is that a Bluetooth service has to be defined in each OnRoute server, *i.e.*, on each vehicle. The Bluetooth service reports which services are defined in its memory. The OnRoute client applications have to discover these services. Because of the limitations of Bluetooth technology and of the actions of the OnRoute servers, declaring services available and the OnRoute client applications discovering said services must be performed properly. This process is one of the major challenges solved by the OnRoute architecture.

The elements of the OnRoute architecture span three domains: the off-board enterprise infrastructure, the nomadic elements (device or service) and the embedded elements (in-vehicle) (see Figure 3). With the OnRoute architecture, when a vehicle reaches a station, a component of the architecture, known as an Information Services Server (ISS), is informed of its availability and its connection address. When the passenger's OnRoute client application seeks available services, it contacts the ISS. The ISS sends the client all of the active services in its database and their connection addresses. The OnRoute client application connects to the OnRoute server in which it is most interested. The OnRoute architecture attains a high degree of service scalability because there is no need for a different service identifier for each OnRoute platform run; they all use the same identifier and report their connection address to the ISS. The time required to search for the services is also reduced because the mobile phones only have to search for a single engine, the ISS. They then connect directly to the line in which they are interested. This arrangement has proved to be fast because the search for devices is slow in devices such as mobile telephones. The more devices there are nearby, the more pronounced is the limitation. The explanation of the Information Services Server, the OnRoute kernel, and the client application will be given below. This description will be composed of two perspectives, that of the operation principles and that of the software development, presenting the main classes, methods and data structures.

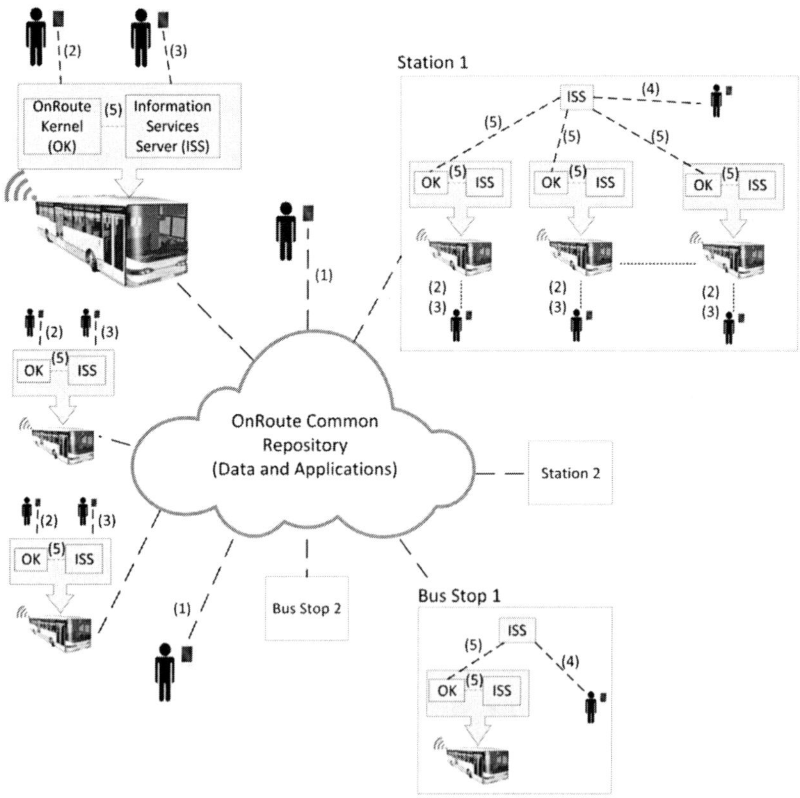

(1) – Transport passengers download the client application.
(2) - The transport passenger accesses a specific information service, the OnRoute guidance assistant.
(3) – The on-board ISS informs the transport passengers about information services available in the vehicle.
(4) – Transport passengers receive information from the ISS (station or bus stop infrastructure) regarding information services available in each vehicles parked at the station.
(5) – The OnRoute kernel communicates the services available/not available in the vehicle to the ISS.

Figure 2. General vision of the OnRoute information services.

4.1. The Information Services Server (ISS)

This entity regulates communication between the OnRoute information services that are run in each vehicle and the client applications. This component of the OnRoute architecture must be run in the central mass transport stations. This approach provides a centralised point of access for the client applications to find what services are available in the station at the time and offers the vehicles a simple way to disseminate their active services. The ISS will also be installed in each of the fleet vehicles to allow a passenger who boards at some point other than the central terminus access to the OnRoute services. The on-board ISS will activate when the vehicle leaves a station, and it will deactivate when it reaches another station, thus using the vehicle location system.

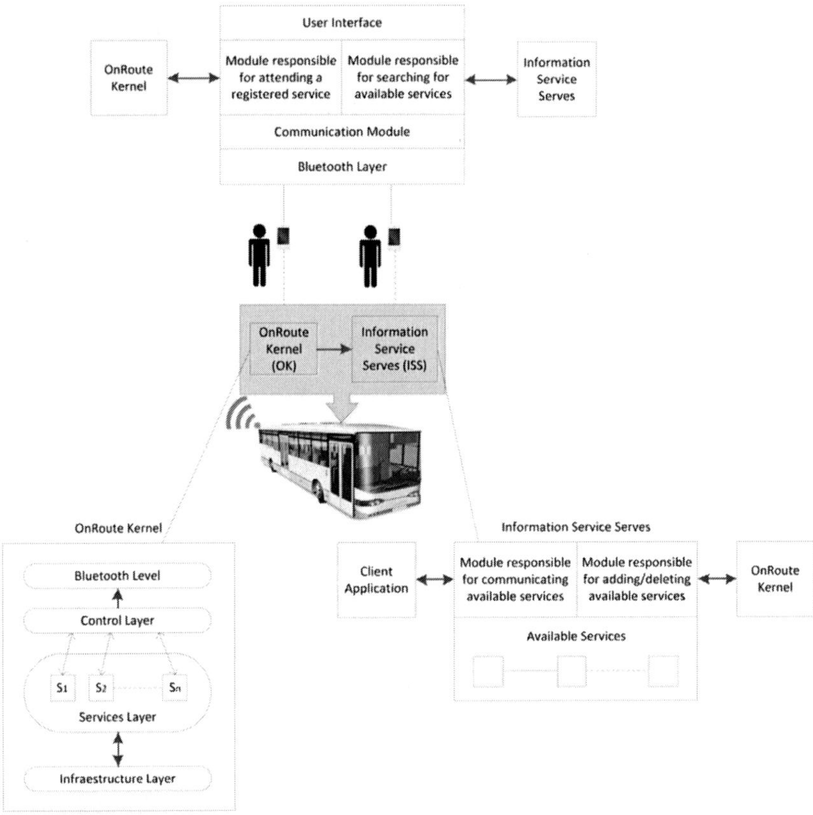

Figure 3. OnRoute general architecture.

From the implementation point of view, the most important data structure of the ISS is the list of active information services. This list is the first thing that is created. Once the list has been created, running the server consists of two concurrent threads: one to deal with the arrival of client applications that need to know which information services are active in the station and the other to listen to the incoming connections from the vehicles to activate or deactivate the information services (see Figure 4).

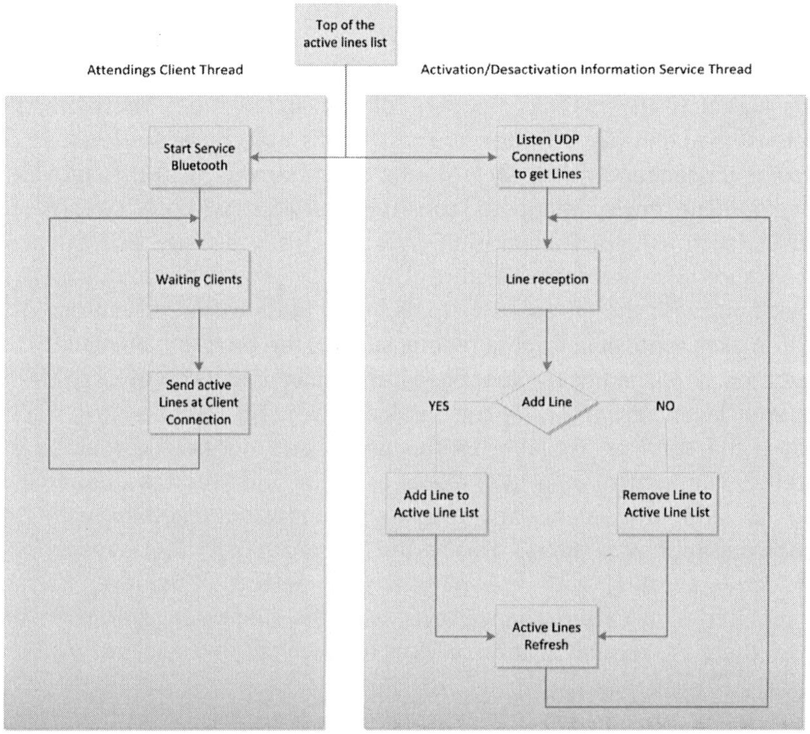

Figure 4. ISS execution flow.

One issue that should be borne in mind with regard to the ISS concerns the Bluetooth technology: the maximum range of a Bluetooth device is 10 metres (for Class 2 Bluetooth devices). The ISS will normally be located in central transport terminals. In most cases, one server will cover an area with a diameter of twenty metres. To ensure service coverage throughout the area, the OnRoute architecture makes a replica of the ISS. This solution is valid because of the way that Bluetooth searches for services. First, a search for devices is performed. Found devices are interrogated in search of the desired service. Because all of the servers will have all of the information, if the client application is within the radius of at least one ISS, it will be able to obtain the necessary data.

4.2. OnRoute Kernel

This kernel is the main component of the architecture. Therefore, the system has been named after it. The kernel's objectives are clear: (1) to provide passengers with services during their journeys and (2) to provide a development environment to construct services without concern for details, such as data transmission. We can clearly distinguish four layers (see Figure 3). The Infrastructure Layer (IL) provides communications operations with the on-board systems in the mass transport vehicles. This layer makes it possible to obtain data, such as the time remaining until the next stop is reached or the exact geographic location. It can be used by the superior layers, especially by the Services Layer. The Services Layer (SL) is where the services available to the passengers during the journey are located. The Control Layer (CL) manages the flow of information between the SL and the client applications (Bluetooth), together with the synchronisation with the ISS. Finally, the Bluetooth Layer (BL) is responsible for communicating with the mobile user devices. This layer can be generalised to a data exchange layer with the client applications. In this layer, there are communication services for each of the services of the SL.

4.2.1. The Services Layer (SL)

Because of the SL, a programmer need only focus on the problem being addressed and can ignore the infrastructure, such as the communications. To this end, operating principles are defined together with a set of methods that follow these principles. All OnRoute information services consist of a programme that follows the run order below:

- **Step 1:** Data initiation. Before anything else, the variables necessary for running the service must be initiated. Certain data will come from outside, while others will be calculated when the service is initiated. This will partially depend on which programmer is developing the OnRoute information service.

- **Step 2:** Service publication. Any information service in this layer informs the higher layers that it is starting to operate. Because it is illogical to have services that do not communicate with each other, this step runs constantly, independent of the programmer's intentions.

- **Step 3:** Running the service logic. For all intents and purposes, this is the programme itself. Code is encapsulated and run here. Methods of the CL will be used to establish a dialogue between the service and the remote applications in the mobile devices.

To ensure implementation and to facilitate the work of the programmers, OnRoute encapsulates this flow into an abstract class, known as OnRouteService. Figure 5 shows a series of methods that are run one after another and have to meet two requirements. The first method establishes that the information service has to have a name. Therefore, the class builder demands it. The publication of the service will be automatic and transparent to the user. The second requirement is that all of the services must be able to run continuously. To attain this performance, the abstract class generates instances that are strings. The Thread abstract class is used, providing an abstract method run where we can place the code that runs the string.

OnRouteService
-name : string -initializerStream[0..*] : byte
+OnRouteService() +run() : void #publish() +setinitializerStream(entrada initializerStream[0..*] : byte) : void +getinitializerStream() : byte +getServiceName() : string #initializeData() #serviceLogic() +registerClient(entrada clientId : int, entrada registryData[0..*] : byte) : void

Figure 5. OnRouteService class.

The attributes of the OnRouteService class are name, the name of the service, and initializerStream, a stream of data that is passed up to higher layers to construct the interface for the client applications. Each programmer must construct this datum in agreement with what is performed in the remote application.

The operations of OnRouteService include a mandatory constructor. The class is created by invoking the constructor, and it must be given the name of the service. The private publish method is responsible for publishing the service. The programmer must invoke the public setinitializerStream method from the method devoted to initialising variables. The method requires a data stream to construct the interface in the client application. If the service requires initialised data, then the iniatilizeDat operation must be invoked, and this operation has to be implemented by the service developer. In the serviceLogic methods, all of the service logic is implemented. Therefore, its code must be specified by the developer. The run method comes from the Thread abstract class. In this class, the three methods that ensure the run flow of an OnRoute service are run in order. Finally, the logic needed to process the registration of a new client application to the service is specified using the registerClient method. If customer registration is not required, its implementation can be left empty.

4.2.2. Bluetooth Layer (BL)

The Bluetooth layer, or transport layer, is the most complex part of the system on all levels, including the technological, design and implementation levels. This layer is the key part because the functionality provided by any service or system depends on it. Situated just above the CL, this layer is merely a specialisation of a client system's communication layer. The layer has been developed with the greatest level of abstraction possible together with the minimum amount of feedback from the other parts of the system. As a result, any change in this layer has no lateral effect on the rest of the system. Even the replacement of the Bluetooth technology with IEEE 802.11 or any other that may be considered advisable, such as ZigBee, must be totally transparent to the OnRoute services.

The Service Distributor is a BL service that expedites the search for OnRoute services and their connection addresses. This component is implemented as a thread that is continuously listening using a fixed Bluetooth address used by the ISS to direct customers to OnRoute. It is important to remember that the Service Distributor will not store the active services. It will have a list of the services, and it will send them to anybody who logs on. The services will be stored in a data structure created for this purpose. The Service Distributor has to be aware of any changes that occur in this structure. To model this behaviour, we will use design patterns. Specifically, we will use the *Observer* pattern in combination with the *Observable* pattern.

Data are sent over Bluetooth in the form of bytes. Therefore, our container has to be converted into a series of bytes that can be interpreted in a certain way at the other end of the connection. There is an intermediate class, ServiceConnectorParser, responsible for acting as an intermediary with the container. This intermediary returns information in the desired format, thereby uncoupling the container from the Service Distributor. If the container implementation changes, the implementation of the parser is changed, which is transparent to the Service Distributor. Similarly, should the Service Distributor's needs change, the container is not affected.

The Bluetooth services are abstracted by the BluetoothOfferedServices class (see Figure 6). This class has the following attributes: dataStream, a chain of bytes that stores the data to be sent to any customer who logs on to the service, and the url, the service address (Bluetooth) to which the client logs on. This class has the constructor method to which the service storage structure has to be sent, a Bluetooth service url and update, which allows the *Observer* pattern to operate. This method is called by the observable datum when its status is modified.

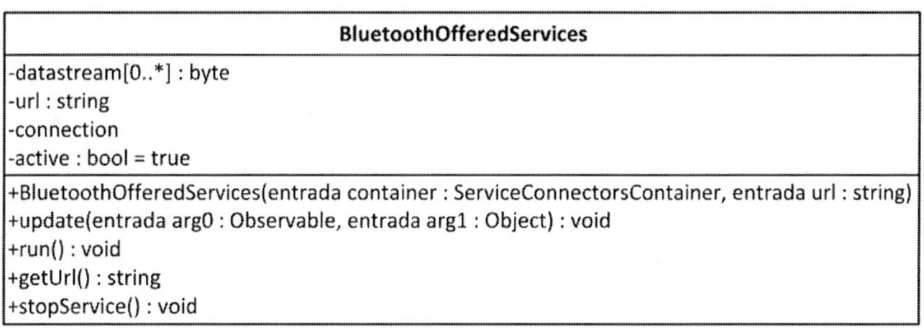

Figure 6. BlueetoothOfferedService.

The Application Initialiser is another component of the architecture placed in this layer. This component allows applications to obtain their initial data. The initialisation data come from the corresponding service in the SL, *i.e.*, they are defined by the developer. It is OnRoute's mission to carry them from one end of the connection to the other. OnRoute only requires that the data be sent to it in an array of bytes. Therefore, any kind of information can be stored and interpreted at the other end of the connection. The packet that the initialiser sends is built as follows:

[+ Registration address +, + Initialisation data +]

Both data are extracted at the other end in a transparent manner for the developer. The registration address will be used once the developer processes the initialisation data.

Another component of the BL is the Applications Registrar. This component implements a registration system that stores the remote addresses of the customers in case it becomes necessary to send them information or initialise a dialogue. As for the registration system, it must be constantly listening. When an incoming connection is made, it stores a remote address to which it can send data or communicate with the client.

With OnRoute, there can be multiple services running at one time. This ability implies that there will be a one-to-one correspondence in the BL. Obviously, there has to be a piece of software that controls everything that happens in the Bluetooth layer and an entry point that allows the control layer to interact. This piece of software is the component of the BL designated as the Bluetooth Controller. Figure 7 shows the interaction between the BL and the OnRoute services.

The Services Container is a fundamental component of this layer. This component stores the services that are going to be activated in the BL and allows them to be managed. We introduced the *Observer—Observable* design pattern, with the Service Distributor as the observer. Therefore, the Service Container is the observable. The Service Container is supported by a hash structure so that searches are conducted in the shortest possible time. This structure is important when there are many OnRoute services available. The methods of this class are (1) the class constructor, which has no parameters and serves only to construct a hash structure, and (2) the addElement, which serves to obtain a Bluetooth service. Whenever an element is successfully inserted, this change is reported to the Service Distributor, refreshing the list of active services. To delete a service, the class has the deleteElement method, which is sent the name of the service to be deleted. Whenever an element is successfully deleted, this change is reported to the Service Distributor, and the list of active services is refreshed. The getConnector method returns the Bluetooth service that coincides with the name sent. Finally, getData returns all of the information contained in the container in an object.

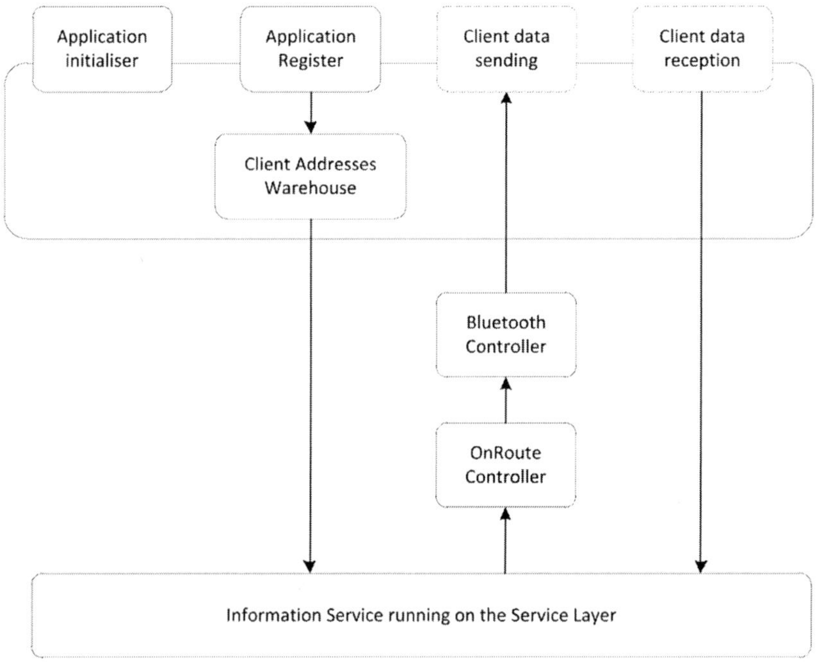

Figure 7. Service Interaction between the Bluetooth layer and the OnRoute services.

4.2.3. The Control Layer (CL)

The CL is responsible for coordinating all the information traffic that occurs in the server. Initialising the first service triggers the activation of the OnRoute Controller, which is the name given to the only CL class. The driver has been created using a singleton pattern, which is designed to restrict the creation of objects belonging to a class. The driver's aim is to guarantee that a class only has one instance and to provide a global point of access.

The use of the *singleton* pattern can be a delicate matter in programmes with multiple running threads. If two running threads attempt to create the first instance simultaneously, only one of them should be able to create the object. To solve this issue, we guarantee the mutual exclusion of the constructor by placing a lock on its point of access. In general, this pattern is applied when this class controls access to a single physical resource or when a certain type of data must be available to all of the other objects of the application.

4.2.4. The Infrastructure Layer (IL)

The purpose of this layer is to enable OnRoute to communicate with the transport vehicle systems. This communication makes it possible to obtain data, such as the vehicle identifier, the identifier and the name of the route, stops on the route and position information. Therefore, it is important to develop a layer that allows communication with all systems that provide information of this kind. Unfortunately, the infrastructures used in transport fleet vehicles vary widely. Each implementation of the infrastructure layer will be different with regard to both the services offered and the methods used to access these services. Note that the data that OnRoute demands from the infrastructure are based on Transmodel [13], which is a European specification that describes a data model for public transport systems whose objective is to facilitate the interoperability of transport networks. This layer can be considered to be a state machine. Each state models the vehicle behaviour at some point of its service from the time it is brought into operation until it is taken out of service. The states are marked by a series of data frames that have to be transported by a medium. We will use the User Datagram Protocol (UDP) because it uses notably little bandwidth.

4.3. The Client Application

The mobile application is based on the MIDlet specification. The implementation is relatively easy, but it is limited. The operation of the mobile application is simple; we have a MIDlet and a set of screens that we use to browse. The operational diagram consists of selecting an item from the screen. This selection triggers a series of actions. Each of these actions is run on a thread that communicates with OnRoute. The thread obtains data, prints them to the screen, and starts over again. The running of any OnRoute client application is governed by a set of general principles. Specifically, each connection with the ISS or an OnRoute system will be made from a new thread, which is sent a different list such that it will display its data on screens. Before the string is invoked, the list is activated by the setcurrent method. Once the thread prints the elements of interaction with the user on the screen, the user selects one and triggers an event that is handled by the commandAction method, and the whole process begins again. Therefore, the data encapsulated will vary between the different run phases of an OnRoute client application.

To send data, the OnRoute ConnectorSender abstract class has been implemented. This class has the connectionUrl stream, and the screen lists and urls as private attributes. The class has the following public operation methods: OnRouteConnectorSender(String connectionUrl, List screen, List urls), run and parse(byte dataToParse[0..*], List screen, List urls). The class is sent the address it has to connect with, and the List object is active on the screen. The object connects and recovers the data. At this moment, the parse method is called by the run method. The parse method is abstract and must be implemented by an OnRoute services developer to enable it to communicate with a service.

For two-way communications between the client application and the service server, a class called CommunicationCenter is provided. This class has the StreamConnectionNotifier type notifier and Boolean type state as private attributes. The public operation methods include stream type CommunicationCenter(String url), haltCenter(), run(), sendData(String url, byte data[0..*]), getUrl() and body(byte data[0..*]. All that is needed is for the user to send the url to which the device is connected. The run method receives the information from OnRoute, and the sendData method allows it to be sent. Finally, every time run receives data, an abstract method, the body, is called to process the data. This abstract method has to be defined by the service programmer.

5. TECHNOLOGY

The software produced using OnRoute can be characterised by its capacity to integrate the surrounding physical and technological environments. Consequently, the software can operate autonomously and spontaneously in different environments. We will have a set of OnRoute services with varying availability and a set of OnRoute client applications that will vary in number. If a passenger is in a station that is busy in terms of both passengers and vehicles, the number can be massive. To attain these functionalities, the OnRoute architecture is based on the architecture of pervasive systems [14].

OnRoute uses Bluetooth technology because it is widely supported by users' mobile devices. The information that is exchanged between two Bluetooth units through a set of slots forms a data package, with each device having a unique 48-bit address based on the IEEE 802.11 standard for WLAN. Two or more Bluetooth units that share a single channel from a Bluetooth network are called a piconet. Although the channels have a bandwidth of 1 Mbit, the capacity is reduced to approximately 10 kbit/s as more users join the network. At most, 8 Bluetooth devices can be part of a *piconet*, assuming one of these devices has the role of the master. To minimise the effect of this limitation, the Bluetooth technology incorporates the concept of a *scatternet*, which consists of two or more integrated *piconets*. Currently, the system architecture assumes the use of piconets in the vehicles. Therefore, up to 7 users can access information services that are provided in each vehicle by the system. *Scatternets* are used at stations so that users can access the information provided by the ISS using their mobile devices.

The OnRoute applications that the passengers run on their mobile devices are developed in J2ME. There are two configurations defined in J2ME: the Connected Limited Device Configuration (CLDC) for devices with processing and memory restrictions and the Connected Device Configuration (CDC) for devices with greater resources. CLDC is used for the development of passenger applications.

6. A PRACTICAL CASE: IMPLEMENTING A GUIDANCE SERVICE

An example of a transport information service is developed in this section, specifically, an on-route guidance service. This service offers a passenger the chance to consult his mobile phone regarding the stops during the journey, subscribe to a destination stop and be advised when the vehicle is about to arrive at that stop. Therefore, the service has to know the name of the route, the stops along the route and the last stop made by the vehicle. All of these data can be obtained from the IL. To begin, we create a new class called InfraestructureService. This class inherits from the OnRouteService abstract class.

First, we define the communications between the service and the IL. This definition is done in the service constructor; we use the PacketTransformer class. This class triggers all of the actions in the IL. We have to send an object from the Connector class, which is responsible for making the network connections with the infrastructure, and the Blackboard class, which is our blackboard object. We then activate the PacketTransformer; therefore, it starts to receive packets from the infrastructure. The packets are processed, and all of the data are initialised. The service is on standby, waiting to receive all of the stops along the route because the mobile application data starter has to be constructed. The source code of this step is as follows:

```
infrastructure = new PacketTransformer(new Connector("UDP", 50000, 1024), blackboard);
infrastructure.start();
while (stopsLinea == null) {
  stopsLinea = infrastructure.getStops();
}
```

In the implementation of the initializeData() method, which is responsible for initialising the data sent to the mobile client application, we recover all of the stops. For each stop, a line of text is generated, which includes the name of the route and its identifier. The source code of this second step is:

```
data = data + "(" +stop.stop.getName() + ":" !+ stop.stop.getCode() + ")";
```

Once the data are generated, it is important that the setinitializerStream(byte[]) method is used to store the data correctly such that it can be used by the higher layers. The source code for this initialisation is

```
this.setinitializerStream(data.getBytes());
```

We maintain a hash structure for the registration of client applications. This makes searches as fast as possible. We use a new ExtendedStop structure in the hash that stores the stops along the route. Moreover, this structure will have an array of whole numbers that will make it possible to register the client. The definition of this structure is presented below:

```
class ExtendedStop {
  public Stop stop;
  public int clients[];
  public int iter = 0;
}
```

We initialise the data in the initializeData() method, using the fact that we obtained the stop data to generate the initialiser of remote applications. The code of this initialisation is

```
container.put(stop.stop.getCode(), stop);
```

The implementation of the registerClient abstract method consists of finding the destination stop in the hash structure, recovering the stop (ExtendedStop structure) and keeping the client identifier in the dedicated array. This is performed as follows:

```
ExtendedStop stoptarget = container.get(registryData);
stoptarget.clients[stoptarget.iter] = clientId;
stoptarget.iter++;
```

Before discussing the service logic, we must consider the operation of the infrastructure layer. The base of this layer is the BlackBoard class, where the data received from the infrastructure are written. This structure is an Observable, and our service will be the *Observer*. Therefore, the update abstract method has to be implemented. Any changes to the data imply a warning to the client because a change means that the stop has changed. The code to get the next stops from the blackboard is

```
synchronized (this) {
  nextstop = blackboard.getnextStop();
  newstate = true;
}
```

The code is executed on a critical section because the variable nextStop controls the logic of the service, and this logic runs on a loop without exit. The loop is iterated only when a change occurs on the blackboard. The next source code illustrates this property:

```
while (true) {
  if (newstate) {
    newstate = false;
    ExtendedStop nextstop = container
     .get(nextstop);
    System.out.println("The next stop is " + nextstop.stop.getName());
    for (int i = 0; i < nextstop.clients.length; i++) {
OnRouteController.getInstance().sendData(this.getName(),nextstop.clients[i],  "The  next
stop is the end of yours trip".getBytes());
}}}
```

7. PERFORMANCE

In this section, we describe the tests performed to assess the validity and performance of OnRoute. Because this report presents a framework for the development of ubiquitous information services for the public transport, the tests have been developed to check (1) the quality of the software produced using its development environment and (2) the performance of the services produced.

The system software was developed in accordance with the Test-Driven Development (TDD) model. This methodology for the development and validation of software runs in two stages. In the first stage, unit tests are developed (Test First Development). In the second state, the code is refactored (Refactoring). The methodology allows us to produce clean code that works. In addition, this methodology enables us to assess the flexibility of the software system that is developed because the system has to be flexible enough to allow it to be tested automatically. Otherwise, TDD would not work. The testing was conducted using a test environment that assumes the TDD methodology, the Java Unit Testing Tools (JUnit). The tests were conducted in accordance with the following guidelines.

In the description of the tests performed, we focus on two critical aspects of the system: the Bluetooth communications and the module of utilities used by the programmers who want to develop information services using OnRoute.

For the Bluetooth communications, the first test conducted was the Bluetooth service (BluetoothService). The purpose of this test was to demonstrate the integrity of the data, ensuring that these were consistent. The next test was to verify that the registration of Bluetooth clients was conducted correctly. The class BluetoothServiceRegister is responsible for this action. The purpose of this class is to register the clients who want to use the service. This class should have a Bluetooth address assigned. The test consisted of passing a service address and verifying that the Bluetooth address assigned was correct.

The tests related to the package of utilities were performed with a critical element: the service container. To verify that the services are inserted correctly in the container, two tests were conducted:

1 To verify that the services are inserted correctly, we generated three services and used the getConnector method to verify that they are equal at the object level. In addition, we verified that the names of the services matched.

2 The second test invoked the method responsible for deleting an element and verified whether the element was still in the container.

With these tests, we guarantee that the data are correct in the most critical parts of the system. However, explicit information services are needed to test the system. Three service prototypes have been developed: a route guidance assistant for travellers, a payment system and an inspection system. In general, any service developed with OnRoute must be evaluated with respect to two aspects: the performance of the Bluetooth services and the time required to provide the information requested by the user. The first aspect can be evaluated independent of the information service. The second aspect depends on the logic of the service and the amount of data handled by the service.

To evaluate the performance of the Bluetooth services, the stages through which any Bluetooth communication passes have been analysed. For a client device, the Bluetooth communication processes include searching for the Bluetooth device used by the information server, obtaining the information services provided by the server, establishing a connection to the server, and opening the connection to the server.

For the server, the stages are specifying the attributes of the Bluetooth service offered, creating the connection for the communication with the customer, and opening the connection for the communication with the client.

The performance of the Bluetooth service is heavily conditioned by the searches that the customers' Bluetooth devices perform. As explained above, the OnRoute architecture offers an element belonging to its “run-time” with the mission of facilitating the search, the ISS. The client devices only have to connect to an ISS to obtain information about the Bluetooth servers available. The Bluetooth clients do not need to perform these costly, time-consuming searches. The performance of this configuration has been analysed in two scenarios: a piconet configuration formed by a maximum of 8 Bluetooth devices and a scatternet configuration, where it is possible to have more than 8 devices. The time required for searching ranged between 10 and 20 s, regardless of the configuration.

Tests were also performed to determine the time required to provide information requested by a user from the information service (response time). We conclude that the main factor affecting the response time is the amount of data communicated. Different amounts of data were requested in a piconet configuration. The transmission of data smaller than 512 B was instantaneous (1 s or less), 1 kB was transmitted in approximately 2 s, 79 kB in approximately 15 s and 300 kB required 2 m. An example of an information service that uses small data packages of 512 B or less is the payment system. In contrast, the inspection system requires data packages ranging in size from 512 B to 1,024 B. Finally, considering these results in the implementation of the route guidance assistant for travellers, the maximum size of the packages of data to be transmitted was limited to 1 kB.

8. CONCLUSIONS AND FUTURE WORK

OnRoute is a framework for developing and running pervasive software that provides public transport passengers with information services. The OnRoute architecture is deployed in different places on the transport network (vehicles, stations and stops). Moreover, this system has a universal vocation, aiming to reach as many users as possible. OnRoute is flexible enough to be installed in any means of transport.

The information systems currently used in public transport networks handle a large quantity of data that is not always available to the user. OnRoute also serves as a bridge between such systems and the passengers. However, OnRoute does not only cover the needs of passengers; it also covers the needs of those who have to construct information services for passengers, *i.e.*, developers. OnRoute offers any interested developer a simple way to create their services, allowing them to focus on the service's functionality and not on the implementation of complex communication systems between users and machines. OnRoute's architecture is understandable and powerful. Because this system's information services are user-friendly, its use does not require significant resources. OnRoute is a multiplatform development tool; it can be installed in Windows, Mac OS or Linux systems. OnRoute has been tested in Fedora and Ubuntu distributions. This platform is fully documented (using Javadoc) and can easily be integrated into general-purpose development frameworks, such as Eclipse.

There are two main challenges that we must face in the system that we have developed. The first challenge is the limitations of the Bluetooth technology that is used to communicate with users' mobile devices. The second is that the applications are J, such as Android or iOS. Because of these limitations, we propose two future studies. First, we should integrate other technologies to perform local communications with mobile devices, such as IEEE 802.11 and ZigBee. This improvement may be easily accomplished because of the modular and layered architecture of OnRoute. The modules have a great degree of abstraction and a minimum level of coupling among different components. Secondly, we must integrate the Android and iOS technology such that the client applications can be executed on Android and iOS devices. To achieve this goal, we will have to update the OnRoute development framework, but it will not be necessary to change the kernel elements, *i.e.*, the ISS and OnRoute kernel that are executed in vehicles.

ACKNOWLEDGMENTS

The authors of this paper wish to express their sincere gratitude to the Transport Authority of Gran Canaria for their support of this work.

REFERENCES

1. Weiser, M. The computer for the 21st century. *IEEE Pervasive Comput. Mob. Ubiquitous Syst.***2002**, *1*, 18–25.

2. Giannopoulos, G.A. The application of information and communication technologies in transport. *Eur. J. Oper. Res.* **2004**, *152*, 302–320.

3. Meier, R.; Harrington, A.; Cahill, V. A Framework for Integrating Existing and Novel Intelligent Transportation Systems. Proceedings of the 8th International IEEE Conference on Intelligent Transportation Systems 2005, Vienna, Austria, 13–16 September 2005; pp. 650–655.

4. Gluli, T.J.; Watson, D.; Prassad, K.V. The last inch at 70 miles per hour. *IEEE Pervasive Comput. Mob. Ubiquitous Syst.* **2006**, *5*, 20–27.

5. Ossowski, S.; Hernández, J.Z. Multi-agent systems for decision support: A case study in the transportation management domain. *Appl. Artif. Intell.* **2004**, *18*, 779–795.

6. Harrington, A.; Cahill, V. Model-Driven Engineering of Planning and Optimisation Algorithms for Pervasive Computing Environments. Proceedings of the 2011 IEEE International Conference on Pervasive Computing and Communications (PerCom), Seattle, WA, USA, 21–25 March 2011; pp. 172–180.

7. López de Ipiña, D.; Vázquez, J.I.; García, D.; Fernández, J.; García, I.; Sainz, D.; Almeida, A. EMI^{2}lets: A reflective framework for enabling AmI. *J. Univ. Comput. Sci.* **2006**, *12*, 297–314.

8. Barretto, S.; Mira, M. Using integration frameworks for developing context-aware application. *Lect. Notes Comput. Sci.* **2004**, *3295*, 100–111.

9. Römer, K.; Schoch, T.; Mattern, F.; Dübendorfer, T. Smart identification frameworks for ubiquitous computing applications. *Wirel. Netw.* **2004**, *10*, 689–700.

10. García-Herranz, M.; Haya, P.; Alamán, X. Towards a ubiquitous end-user programming system for smart spaces. *J. Univ. Comput. Sci.* **2010**, *16*, 1633–1649.

11. Kindberg, T.; Fox, A. System software for ubiquitous computing. *IEEE Pervasive Comput. Mob. Ubiquitous Syst.* **2002**, *1*, 70–81.

12. Hervás, R.; Bravo, J.; Fontecha, J. A context model based on ontological languages: A proposal for information visualization. *J. Univ. Comput. Sci.* **2010**, *16*, 1539–1555.

13. European Committee Standardization. *Reference Data Model for Public Transport*; Technical Report CEN TC278; European Committee Standardization: Brussels, Belgium, 2005.

14. Saha, D.; Ukherjee, A. Pervasive computing. A paradigm for 21st century. *Computer* **2003**, *36*, 25–31.

Chapter 5

Bluetooth in Intelligent Transportation Systems: A Survey

M. R. Friesen[1], R. D. McLeod[2]

[1]Design Engineering, University of Manitoba, Winnipeg, MB R3T 5V6, Canada

[2]Electrical and Computer Engineering, University of Manitoba, Winnipeg, MB R3T 5V6, Canada

ABSTRACT

The rise of Bluetooth-equipped devices in personal consumer electronics and in in-car systems has revealed the potential to develop Bluetooth sensor systems for applications in intelligent transportation systems. These applications may include measurements of traffic presence, density, and flow, as well as longitudinal and comparative traffic analysis. A basic Bluetooth sensor system for traffic monitoring consists of a Bluetooth probe device (s) that scans for other Bluetooth-enabled device (s) within its radio proximity, and then stores the data for future analysis and use. The scanned devices are typically on-board vehicular electronics and consumer devices carried by the driver and/or passengers which use Bluetooth communications, and which then reasonably proxy for the vehicle itself. This paper surveys the scope and evolution of these systems, with system attributes and design decisions illustrated via a reference design. The work provides motivation for continued development of non-invasive systems that leverage the existing communication infrastructure and consumer devices that incorporate short range communication technology like Bluetooth.

KEYWORDS

Bluetooth sensing, Traffic monitoring, Intelligent transportation systems

1. INTRODUCTION

Intelligent traffic systems (ITS) hold the promise to improve roadway congestion and transportation infrastructure management by capitalizing on information derived from traffic monitoring. The increasing requirement and public expectation for accurate vehicular traffic information to manage traffic flows has precipitated the deployment of large scale traffic monitoring infrastructures. Typically, this has included the use of inductive loop detectors, microwave sensors and relatively expensive video cameras.

On-board vehicular electronic devices as well as consumer electronic devices are emerging as an alternative traffic sensing modality to complement the existing traffic monitoring and management infrastructure. This evolving infrastructure provides has the benefit of providing cost-effective, real time traffic data by leveraging existing telecommunication infrastructure such as the cellular phone network.

This paper reviews the application of Bluetooth sensing in relation to ITS. It is a field that has seen very rapid evolution in the past 10 years, and will undoubtedly continue to evolve rapidly. A survey of the current state of the art can serve as a point of reference for both future and past works. In this paper, wireless sensor networks based on Bluetooth sensing are presented as a practical means of collecting a statistical representation of traffic density and flow. A basic system configuration consists of a Bluetooth probe device (s) that scans for other Bluetooth-enabled device (s) within its radio proximity, and then stores or forwards the data for future analysis and use. The scanned devices are typically on-board vehicular electronics and consumer devices carried by the driver and/or passengers which use Bluetooth communications, which reasonably proxy for the vehicle itself.

Thus, the data provide the information needed to extract a reasonable approximation of traffic presence, density, and flows.

The work reported here is contextualized within many of the uses initially suggested under the IntelliDrive initiative [1] that are oriented towards improving mobility within surface transportation systems. However, advanced applications are not practical to deploy on scale without leveraging implementations based on networked and web services and evolving internet technologies. Currently, web service realization for ITS applications is promising, since a uniform middleware can be achieved while utilizing the underlying network infrastructure [2]. The applications reviewed in this paper demonstrate this integration.

Early vehicular telematic applications were user-centric, whereas innovative applications that are now within reach often combine crowdsourced information with the objective of data collection and analysis for statistical reliability and generalizability [3]. These newer class applications are cohort-centric rather than individual- or user-centric. An early reference to the use of crowdsourcing for ITS [4] pursued the idea via a Smartphone app than from a wireless sensor network. These applications often rely on inferencing from GPS-equipped probes or floating cars, with the intent to capture the behaviours of a statistically significant portion of vehicles, such that meaningful inferences can be made and potentially generalized to the entire population of vehicles [5], [6]. Being statistical in nature, in some cases, only a very small amount of floating car or probe data are required to infer significant events such as congestion build-up or dissolution [7]. In addition to the work relative to estimating traffic from GPS-enabled probe devices and map services, an alternative includes the possibility of using drivers' cellphone trajectories as directly as proxies for vehicles, without an intermediary Smartphone app, as could be provided from a mobile cellular service provider [8–10]. Cellphone trajectories are typically coarser-grained in both space and time but can be overlaid on a traffic grid and allow for some degree of traffic flow inferencing [11].

This survey is directly related to the use of Bluetooth transceivers to crowdsource data from a statistically significant sample of vehicles, where the data can be analyzed for traffic reliable inferencing. Bluetooth is a short-range, wireless telecommunications standard that defines how mobile phones, computers, personal digital assistants, car radios, GPS units, and other digital devices can be easily interconnected. A prototypical example of the technology configuration is the interconnection of a driver's or passengers' mobile phone to a wireless earpiece of vehicle audio system for hands-free operation while driving.

The remainder of this paper is organized as follows. Section 2 provides a background of Bluetooth in Intelligent Transportation Systems (ITS) from some of the earliest academic references in the research area to the present. Section 3 discusses Bluetooth relative to the methods and data that can be readily obtained in a non-intrusive manner in an ITS context. Section 4 overviews the design considerations within a Bluetooth configuration for ITS applications, and Section 5 provides a provides a an illustrative reference design which encompasses many of the phenotypes of a typical Bluetooth sensor network for ITS and illustrates the ease with which data can be collected, stored and presented.

2. A SURVEY OF BLUETOOTH IN ITS

The earliest references to using Bluetooth for tracking purposes were typically unrelated to vehicular traffic and ITS. Within overall objectives of safety and flow monitoring, early examples included Bluetooth tracking systems to track of children at a zoo [12] and students at a University [13]. The realization that wireless sensor networks generally may play a significant role in traffic monitoring emerged in the literature in the mid-2000s [14] without explicitly mentioning Bluetooth.

Bluetooth may now appear to be an obvious methodology for non-intrusive traffic detection and estimation; however, industry reports and tests as late as 2010 evaluated various traffic sensors without considering Bluetooth as an option [15]. In a 2010 study, the authors almost apologetically implied that limiting their system to Bluetooth networks was a potential limitation, and indicated that their work could be applied to Wi-Fi devices as well [16]. The field since then has borne out that the Bluetooth devices serve as better proxies for vehicles than Wi-Fi devices would have been.

Another concurrent stream of research has been associated with the role of GPS in traffic monitoring relative to estimating travel times, vehicle density, and vehicle flows. In a more ambitious program where data was collected from GPS-enabled cellular devices, the work indicated that a 2-3 % penetration of GPS enabled cell phones in the driver population was sufficient to provide accurate measurement of the velocity of traffic flow [6]. Although the percentage of GPS enabled cell phones is clearly above the 2-3 % of the population, it is also a requirement that the GPS unit be physically on and driver data voluntarily be provided (either implicitly or explicitly). In comparison to Bluetooth, the only requirement is that the probed device has its Bluetooth enabled or discoverable, which is more often the case than having GPS enabled as well.

The potential of Bluetooth in traffic monitoring began to appear in the academic literature near 2010 [16, 17] although a small number of early field trials by local government transportation departments and agencies date to as early as 2008 [18] and 2010 [19]. Another early reference to Bluetooth sensing in a 2009 academic thesis emphasized optimal sensor location as opposed to data collection [20]. These were some of the first publications to consider Bluetooth as a means of collecting data for traffic monitoring and ITS management. A potential exception may be a 2004 reference to Bluetooth for ITS where Bluetooth is considered as a means of inter-vehicle communication as opposed the use of Bluetooth as a traffic monitoring sensor [21].

The reliability of Bluetooth sensor systems for ITS depends on widespread Bluetooth-enabled device penetration. In a study of traffic through the Limfjord Tunnel, Bluetooth penetration was estimated at between 27 and 29 % [22]. This appears to be considerably higher than many other reports at or about the same time frame. In a longitudinal study, Bluetooth penetration was seen to increase dramatically with the number of unique MAC addresses seen increasing over a year by 26 % [22]. Rationale for the increased penetration was assumed to be the popularity of GPS units combined with an increasing number of vehicles with built-in Bluetooth. This reported level of Bluetooth penetration augers well for the continued investigation into these technologies and increased levels of statistical confidence as an increasing number of Bluetooth devices become discoverable.

Most often, one thinks of the Bluetooth-enabled cellular phone as a foundational component in a Bluetooth sensor system for ITS. This premise is supported by cellular penetration rates in Canada and the U.S as well as elsewhere. In the first quarter of 2012, Canada had 28 million cellular subscribers in total, which is a penetration rate of up to 80 % if no duplicate devices to a single subscriber are considered. While not all cellular phones are Bluetooth-capable, Nielson reports Smartphone penetration of up to 64 % of mobile phone owners in the U.S. by August 2013 [23].

In one of the earliest references, researchers in 2008 envisioned the basic system components of a Bluetooth sensor system for ITS that have since evolved, stating that "one could easily imagine a battery-powered, Bluetooth enabled, smart cell phone in a plastic case chained to the side of the road to collect much more substantive travel time estimates over 24 h or 7 d to much more precisely characterize operational characteristics of either a signalized corridor or a construction work zone. Those data might be logged for later download" [18].

In general, these early studies typically focused on applications to vehicle travel times estimates (including travel time delays due to roadway obstructions) and origin–destination estimates on urban arterials and freeways. Potential applications to network analysis (shortest path), congestion reporting, bicycle and pedestrian travel times, and before-after studies were also speculated [19]. Other research interests included the quality of data produced by the Bluetooth detection of mobile devices for applications to travel time forecasting and estimates of time-dependent origin–destination matrices within an Advanced Traffic Information System (ATIS) that supplies information to drivers [24]. More recent studies likewise investigate the role of Bluetooth sensors to estimate travel times [25, 26] and vehicle velocities [27].

Privacy considerations were also foreseen early on, with the recommendation that organizations that implement Bluetooth-based tracking for ITS applications develop practices that encrypt and preclude storing MAC addresses for more than a few hours [18].

Early studies also observed asymmetry of traffic data collected in opposite directions (e.g. westbound vs. eastbound) and attributed this to antenna position. Others extended this line of investigation, characterizing antenna patterns and observing that an omnidirectional radiation pattern is most suitable for Bluetooth data collection [28]. This is not unexpected, as it reduces the complexity of antenna placement and analysis when implementing a larger system of Bluetooth sensors.

In one such study of antenna characterization relative to travel time data collection using Bluetooth, the proportion of unique Bluetooth MAC addresses read relative to the known total traffic flow was reported to be 10 % [28]. Although published in 2013, these data appear to have been collected in 2011 and as such, they represent a significant increase in the Bluetooth MAC address reads reported just three years earlier at 1 % [18]. This improvement is likely due to both higher gain antennas developed combined with the precipitous increase in the number of Bluetooth devices in vehicles over that same time period. Currently, the collecting of CoD information is largely absent in the available literature.

Furthermore, the role that Received Signal Strength Indicator (RSSI) may play in the Bluetooth data collection process in ITS arose as an explicit research question around 2011 [29], with little other apparent work on investigating RSSI as a Bluetooth data source specifically for vehicular traffic. One of the few studies available explores using RSSI as means of estimating distance from the scanning point and using Class of Device information as a potential means of differentiating pedestrians from vehicles [30].

System calibration and validation studies are becoming more rigorous as more Bluetooth sensor systems for ITS are deployed. Opportunities exist to calibrate the data with existing traffic measurement devices, including loop-detectors, mechanical counters, travel survey data of various forms, and increasingly, data inferred from cellphone trajectories. In one novel study, Bluetooth sensor data has been compared with automatic license plate recognition (ALPR) systems [17]. If available, ALPR provides as nice alternative for validation as both ALPR and Bluetooth sensors also label the data, in contrast to loop-detectors, counters, and survey data. Systems have been developed that employ a variety of sensors and their fusion in scaling up systems [31], such as the fusion of loop based detection with Bluetooth-enabled devices [32]. While Bluetooth devices appear to have been limited to defined probe vehicles resulting in a low quantity of Bluetooth data, the work demonstrated opportunities in multi-sensor data fusion using Bluetooth data in conjunction with auxiliary measurements. Another instructive example of multi-system calibration potential is with the Anonymous Wireless Address Matching (AWAM) proof of concept demonstration with the City of Houston on an urban arterial [33]. Travel times and speeds collected on identical roadway segments using a probe-based toll tag (AVI) system and the AWAM system were compared with excellent correlation.

While large-scale deployment of commercial Bluetooth traffic monitoring is still in its infancy, a number of pilot programs of various scales are being deployed. In Clark County, a pilot program costing $540,000 has approximately 20 Bluetooth probes installed collecting data along the Andresen corridor which experiences relatively high traffic, in an effort to determine whether the system can provide the information that traffic engineers need [34]. The study reports the system reading 3-5 % percent of vehicular traffic via Bluetooth MAC addresses, and they have recognized this to be sufficient to provide information on traffic flow. According to the authors, sufficient data and analysis is anticipated by early 2014 for agencies to use the outcomes to adjust traffic signal timing. Similar pilot programs are ongoing in many other countries [35–37]. While many pilot programs have focused on traffic flow on corridors and arterials, there are other Bluetooth scanning applications that are associated with work zone diversions [19, 38, 39]. At present, the majority of pilot programs are related to travel time informatics and Bluetooth data assessment towards building evidence-based cases for traffic signal retiming [40].

One of the earliest commercial system that used Bluetooth for vehicle identification for travel time estimation appears to be BLIDS [41] (http://www.blids.cc/). BLIDS was introduced early 2008 with over 50 systems deployed primarily along corridors. Traffax Inc. (http://www.traffaxinc.com/) is also one of the early commercial vendors of Bluetooth traffic monitoring systems, with the system known by the trade name BluFax and introduced in 2009. Traffax has a patent application pending (CA2711278 A1) which claims a priority date of January 2008, (provisional filing) but may come under some challenges as the system of [18] was already deployed at that time. Blipsystems (http://www.blipsystems.com/home/) supplies a commercial solution to Bluetooth tracking and traffic monitoring, with their product is denoted BlipTrack. Another commercial system is TrafficCast (http://trafficcast.com/) with a product denoted BlueTOAD. Iteris (http://www.iteris.com/) also has similar product denoted Vantage Velocity for capturing Bluetooth MAC data but does not appear to have integrated cellular connectivity. In these commercial systems, most if not all of the components discussed in this survey are included as product offerings. Somewhere in-between commercial systems and academic prototypes, various transportation institutes and agencies are leveraging intellectual property developed at or in conjunction with universities. An excellent example is the Texas A&M Transportation Institute [37, 42].

With the observed data gathering potential of Bluetooth for ITS applications combined with the emergence of a 'big data' culture, numerous references to systems and implementations have appeared from industry, governments, and academia. This diversity of approaches, motivations, and originators of the research lends credibility to the expanding role that Bluetooth sensing can play in ITS. In spite of this, there continues to be a need to explore system integration and validation for the various combinations of system configurations that can be envisioned.

3. BLUETOOTH TECHNOLOGY

Bluetooth is one of several available wireless technologies that may be employed to assist in resolving location extracted from a consumer electronic device (Table I). This survey focuses on classical Bluetooth 2.0 with a range that is well-tailored for monitoring or detecting devices residing in or integrated into vehicles, such as Smartphones, Bluetooth earpieces and car audio. NFC and BLE 4.0 are more recent market entries with emphasis on low power and more personal communication or very body-centric networking. WiFi and cellular are intended for wider area networking.

Table I. Various wireless proximities

Technology	Range
NFC (RFID)	~ 20 cm
BLE 4.0	<25 m (estimated)
Bluetooth	1-100 (class dependent)
WiFi	5-100 m (typical)
Cellular	km+(cell sector –typical)

In addition to Table I, there is a variety of wireless networking technologies developed over the IEEE 802.15.4 protocol such as ZigBee or XBee. These and similar are not considered here as they have not established a dominant presence in consumer devices such as smartphones. However, they are technologies to stay aware of as automobile manufacturers incorporate greater degrees of low cost wireless technologies into product lines. Relative to the basic functionality of inferencing vehicle presence by scavenging radio signals, ZigBee could provide an alternative to Bluetooth, although at this time is difficult to foresee ZigBee to be as pervasive as Bluetooth.

In general, Bluetooth scanning requires a device to probe the local wireless environment and detect the proximity of Bluetooth radios. In the ITS context, the proximate Bluetooth radios (typically, drivers' or passengers' Smartphones, earpieces, or on-board car audio) detected by the probe device serve as proxies for vehicles. The objective of the probe devices (aka probes) is to collect information on vehicle presence (detection by one probe) and vehicle trajectory information (detection by multiple probes in sequence) via Bluetooth device discovery, and then transmit this information via either an intermediate wireless tier or directly over a cellular network to a web service or backend server. The communication protocols between probes and backend servers are typically based on the TCP/IP protocol stack, leveraging not only the physical communication infrastructure but also the highly developed Internet IP infrastructure.

The data collected from a Bluetooth scan can be fairly detailed, including a detected Medium Access Control (MAC) address, Class of Device (CoD) information, as well as metadata from the manufacturer or as specified by the device owner (Fig. 1). While getting all of this information from a device and storing it could lead to security and privacy concerns, the MAC address, the primary method of identifying a Bluetooth device, can easily be anonymized in a fashion such that the device can be uniquely identifiable from other devices while preserving a user's identity to a certain degree.

The Bluetooth address itself is a unique 48bit device identifier, where the first 3 bytes of the address are assigned to a specific device manufacturer by the IEEE (www.ieee.org/), and the last 3 bytes are freely allocated by the manufacturer. Even if the manufacturer of a device is known, the number of possible Bluetooth addresses is immediately limited to 16,777,216. As well, since only devices that are in discoverable mode can be detected by the scan, a user can simply turn their device off (non-discoverable) to avoid being detected. There are software tools available which allow brute-force discovery of non-discoverable devices, an early example of which is RedFang, but this is usually too complex for a minimal hardware configuration and not necessary for data collection purposes associated with ITS. It is also possible to burrow deeper into Bluetooth connections to provide connection-based tracking for fine granular or building-wide device tracking [43].

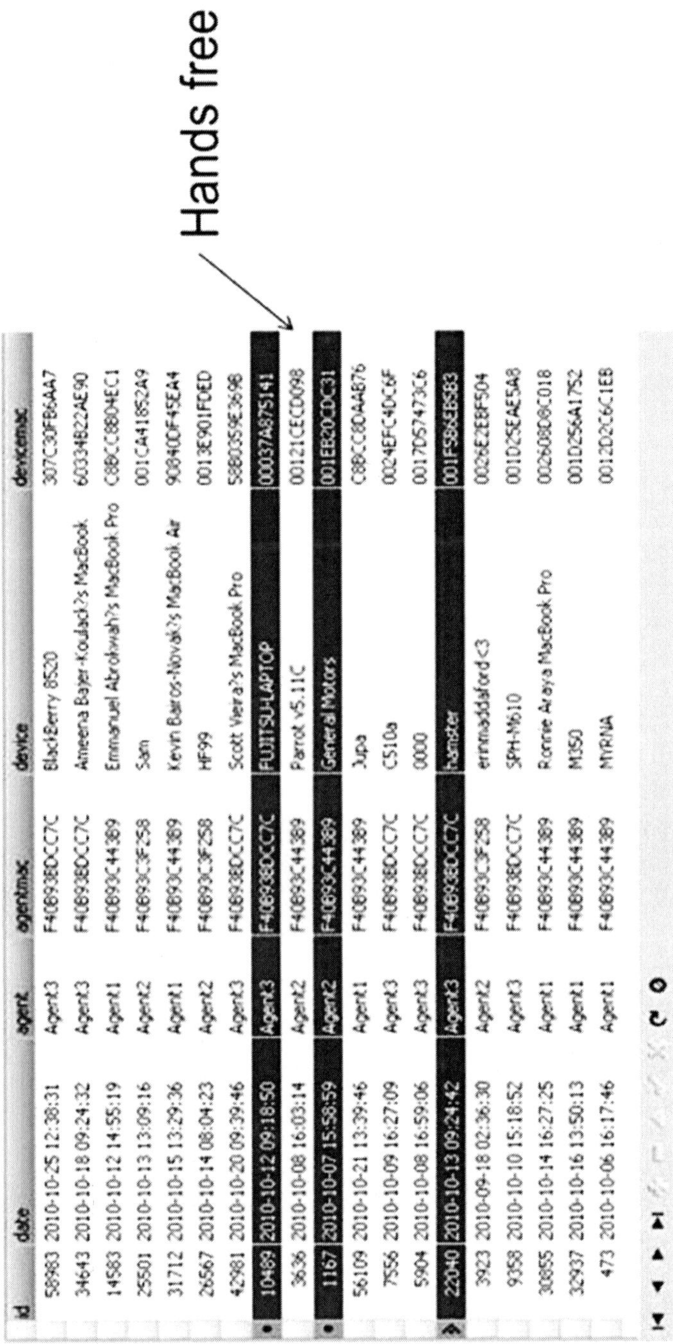

id	date	agent	agentmac	device	devicemac
58983	2010-10-25 12:38:31	Agent3	F40B93BDCC7C	BlackBerry 8520	307C30FB6AA7
34643	2010-10-18 09:24:32	Agent3	F40B93BDCC7C	Ameena Bajer-Koulack?s MacBook	60334B22AE90
14583	2010-10-12 14:55:19	Agent1	F40B93C44389	Emmanuel Abrokwah?s MacBook Pro	C8BCC8B04EC1
25501	2010-10-13 13:09:16	Agent2	F40B93C3F258	Sam	001CA41852A9
31712	2010-10-15 13:29:36	Agent1	F40B93C44389	Kevin Bairos-Novak?s MacBook Air	90840DF45EA4
26567	2010-10-14 08:04:23	Agent2	F40B93C3F258	HF99	0013E901FDED
42981	2010-10-20 09:39:46	Agent3	F40B93BDCC7C	Scott Vieira?s MacBook Pro	58B0359E369B
10489	2010-10-12 09:18:50	Agent3	F40B93BDCC7C	FUJITSU-LAPTOP	00037A875141
3636	2010-10-08 16:03:14	Agent2	F40B93C44389	Parrot v5.11C	00121CECD098
1167	2010-10-07 15:58:59	Agent2	F40B93C44389	General Motors	001EB20CDC31
56109	2010-10-21 13:39:46	Agent1	F40B93C44389	Jupa	C8BCC8DAAB76
7556	2010-10-09 16:27:09	Agent3	F40B93BDCC7C	C510a	0024EFC4DC6F
5904	2010-10-08 16:59:06	Agent3	F40B93BDCC7C	0000	0017D57473C6
22040	2010-10-13 09:24:42	Agent3	F40B93BDCC7C	hamster	001F5B6EB5B3
3923	2010-09-18 02:36:30	Agent2	F40B93C3F258	erinmaddaford<3	0026E2EBF504
9358	2010-10-10 15:18:52	Agent3	F40B93BDCC7C	SPH-M610	001D25EAE5A8
30855	2010-10-14 16:27:25	Agent1	F40B93C44389	Ronnie Araya MacBook Pro	002608D8C018
32937	2010-10-16 13:50:13	Agent1	F40B93C44389	M350	001D256A1752
473	2010-10-06 16:17:46	Agent1	F40B93C44389	MYRNA	0012D2C6C1EB

Figure 1. Simple Bluetooth Probe Scan Data

It is desirable, but not necessary, to employ probes in such a way that the system is scalable. This is one of the most unique features of a Bluetooth data collection system. An organization can start with a very limited and even a portable system and easily scale it to meet increased sensing demands. In this regard, the backend data collection, storage and analysis hardware are conservative. Modern web servers and connectivity are more than sufficient to collect the volume of data that could be collected even from a large number of Bluetooth probes. The scalability issue has essentially been further resolved by existing cellular and internet infrastructure, web servers, and data services collectively being developed as a Service-Oriented Architecture (SOA) [44, 45].

The primary equipment used as the Bluetooth probe device must have the ability to be set for device discovery. Several of the existing Bluetooth modules libraries are designed to detect eight devices per inquiry, which is insufficient for most vehicular traffic applications. The limitation of eight appears to be a consequence of the anticipated use cases of Bluetooth enabled consumer devices, where supporting eight connections is likely more than sufficient. In effect, many existing Bluetooth modules are designed for connecting with the Bluetooth devices it discovers, whereas for ITS Bluetooth data collection, the primary requirement is only for devices to be detected. During traffic congestion periods, it is desirable to discover as many surrounding devices as possible, but a connection to the surrounding devices is not required. The important aspect is to ensure that the selected Bluetooth probe modules can discover as many devices as possible. In the reference design (Section 5), up to 250 unique devices per inquiry were detectable and hence recordable. This detection capability is not a restriction of the standard but rather particular implementations. For example, a detailed description of the Bluetooth discovery protocol that simulates the detection of 15 devices within a few seconds is available in the literature [46].

4. BLUETOOTH SENSOR SYSTEM: ATTRIBUTES & DESIGN DECISIONS

When designing a Bluetooth sensor system for ITS applications, there are choices in system attributes that become design decisions unique to the context and objectives of the system in deployment (Fig. 2). The basic configuration requires the designer to decide what type of probe device (s) will be used, how many probe device (s) are required, and where and how they will be located and fixed in the environment.

- Commercial/Prototype: In academia, a prototype system can often be easily assembled for several thousand dollars and provide entry-level data for exploring Bluetooth device detection. A commercial-grade installation of similar scope would likely be an order of magnitude greater in cost. Intermediate-grade systems are also available, for example [30].

- Fixed/Portable: A fixed system implies a permanent or semi-permanent installation, whereas a portable system implies a system set-up or tear-down time of several hours maximum, and a deployment period measured in days, weeks, or months. In the case of a portable system, consideration has to be application domain, and a portable system offers the convenience of being more easily redeployed. A portable system typically implies a storage battery and possibly a battery charging system, such as solar. A fixed system is typically more robust but greater attention in sensor placement is required as the initial placement decisions may not be easily changeable. Fixed system may also rely on battery power, although a permanent power supply may be cost-effective as well. In both fixed and portable systems, GPS positioning is required. Both fixed and portable systems may be online or offline

- Online/Offline: An on-line system has the potential for real-time data collection (and potentially, analysis), by data being stored in the probe for only very short periods of time and backhauled at regular intervals over a wireless connection to a dedicated server. In an offline system, a requirement exists for much more significant data storage and relatively simple retrieval. A minimum requirement would be for the probes to write data to an onboard SD card or similar, and the manual data retrieval protocol to be planned. On-board data storage is typically Flash-based and very robust.

- Wireline power/battery/solar: The choice of power to the probe devices depends largely on the intended application. A portable application will most probably be battery powered, and depending on it intended duration and the environment, it may also be equipped with a solar charging system. Wireline power may be an economical alternative in a fixed system, although a fixed system may also run on battery power.

- Networking: Networking considerations are typically limited to online systems, where the designer must consider means of data transport. Two of the more likely networking technologies are cellular (e.g. GSM), or WiFi. The consideration in selecting WiFi would be to ensure adequate coverage over a wide area.

- Tiered wireless configuration: A tiered wireless sensor network is a very common architecture for sensor networks (e.g. [30] and the reference design in Section 5). In a system with multiple probe devices, a design decision must be made whether each probe will be equipped with its own WiFi or GSM/GPRS module, allowing direct communication from each probe to the back-end server. The alternate would be to implement a middle wireless tier, adding some complexity to the system. An example, of a multi- tiered wireless implementation is presented in Section 5. It is these authors opinion however, that a tired wireless system should be avoided and direct communication to the probes be supported.

- RSSI capable: Only a very limited number of published works incorporated RSSI for traffic monitoring in an ITS context [29,30, 47]. At time of writing, RSSI data does not play a significant role in Bluetooth systems for ITS applications, although the rapid evolution of this area in general may illuminate the potential and utility of RSSI data within a few years.

- Bidirectional: A bidirectional system may be able to proactively communicate over Bluetooth to 'subscribers' or discovered devices, for example by providing traffic alerts. For example, a potential business model would allow users to purchase low-cost devices solely for the purpose of being tracked [18], with a subsequent feature of being able to backcast from the ITS to the device. This device may be as redundant as a Smartphone running a suitable app. The opportunity to also log a subscriber's OBD-II data automatically would also require bidirectional communication via an established connection with a simple Bluetooth OBD-II dongle.

- Remote monitoring and/or control: The Bluetooth probes should be chosen to be remotely monitored and preferably also configurable. The physical environment of installation sites (heavy traffic areas, exposure to all weather conditions) may make on-site monitoring and configuration both uncomfortable and potential dangerous. Remote monitoring allows for early detection of malfunctioning probes and other inconsistencies. Remote configuration can range from configuring the sampling rate and/or sleeping sensors when not required, which becomes critical for battery powered units. The implication of remote monitoring and configuration is that wireless access via WiFi or cellular is available to the probes. .

- Cross validation/calibrating: As a Bluetooth sensor is sampling a proportion of the by-proxy vehicle population, by definition considerable emphasis has to be placed of validating and calibrating. A Bluetooth sensor network is relatively easy to install and provides considerable opportunity for sensor fusion. At minimum, designers must consider Bluetooth radio ranges of the probe (s) relative to the sampling area and clock synchronization between multiple probes. In data analysis, considerations include but are not limited to the ability to handle multiple MAC address reads from a given probe sensor that represent different vehicles as well as multiple reads of the same vehicle (e.g. a vehicle stopped at a traffic signal for several sampling periods), simultaneous MAC address reads from two or more probe sensors, one or more MAC address reads of a single MAC address by multiple sensors either simultaneously or in sequence, and MAC address reads of devices that are not necessarily sourced from a vehicle. The fact that the data is labeled can be used to provide some level of differentiation. There will always be some degree of uncertainty; for example, the data analysis is unlikely to be able to definitely differentiate a single vehicle from public transportation (e.g. a bus with 40 passengers and multiple Bluetooth reads). Managing uncertainty is one of the more academic aspects associated with Bluetooth traffic monitoring and promises to be a rich area of research [48].

- Security: By virtue of the fact that all wireless communication devices need to signal to some degree in plain sight, security will always be an issue. It is not possible to alter this fact, as communicating devices need a standard means of identifying one another.

Bluetooth scanning for ITS applications such as simple traffic flow will become contextualized within big data concepts, in which the opportunities for exploring and exploiting the rich suite of data such systems are capable of generating is a significant research area in its own right. Examples include generating trajectories from uncertainties in measurement and detection [49] using techniques like Hidden Markov Models. Similar efforts will produce forecasting models that use massive amounts of real-time Bluetooth device data.

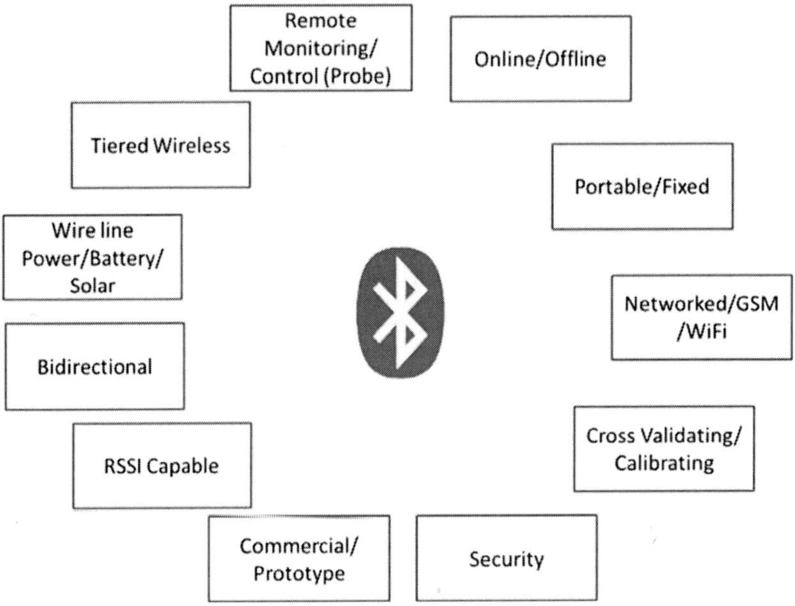

Figure 2. Bluetooth traffic monitoring design decisions

5. A REFERENCE DESIGN

This section presents a reference design as an example of one combination of many of the system attributes and design choices overviewed above. The system used multiple Bluetooth transceivers consisting of one master node or access point and multiple sensor probes around a major intersection in a Winnipeg, Canada, during winter 2013 (Fig. 3) [50]. The multiple probes collected vehicle presence information (detection by one probe) and vehicle trajectory information (detection by multiple probes in sequence) via Bluetooth device discovery, and then transmitted this information to the master node via the 802.15.4 protocol.

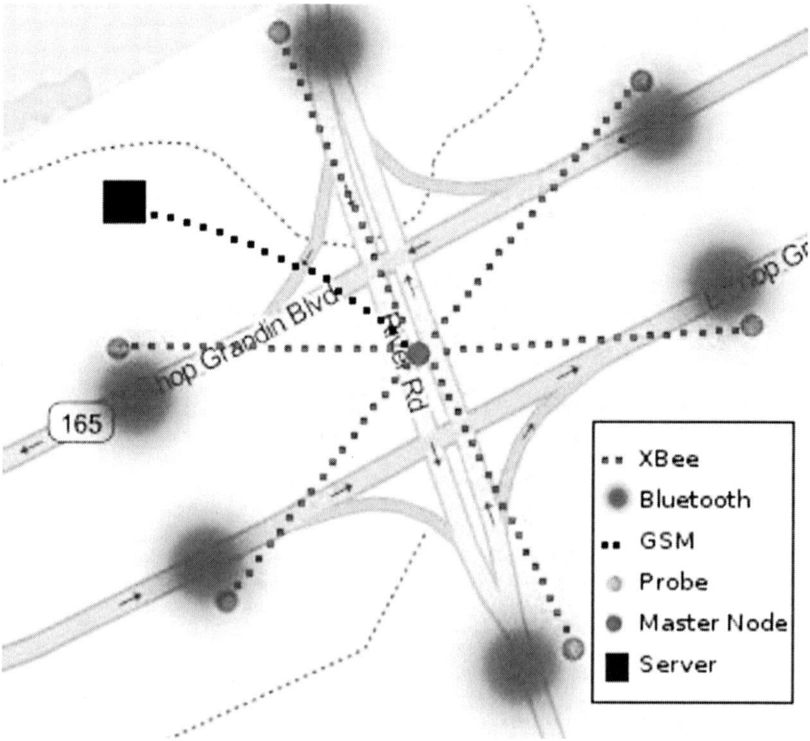

Figure 3. High level system overview

5.1 Architecture

The basic system architecture is that of a Bluetooth sensor network, interconnected with an XBee/802.15.4 middle tier to the master node, and then a GSM wireless backhaul tether to a web data collection and processing web portal. XBee Pro was selected as the middle tier wireless networking technology as it offers a low power solution with sufficient range for the interconnection between the master node and sensor probes. GSM was selected as the cellular tier as a means of aggregating and forwarding data collected to a web server for processing and display. The data sent to the central server displayed the current traffic density and average velocity of vehicles at the intersection on a web front end (including mobile website) at five-minute intervals.

5.2 Probe design

The probe design uses an Arduino Uno development board, which utilizes the ATMega328 microcontroller, an XBee module and a Bluetooth Pro module, both of which connect to the Arduino module (Fig. 4).

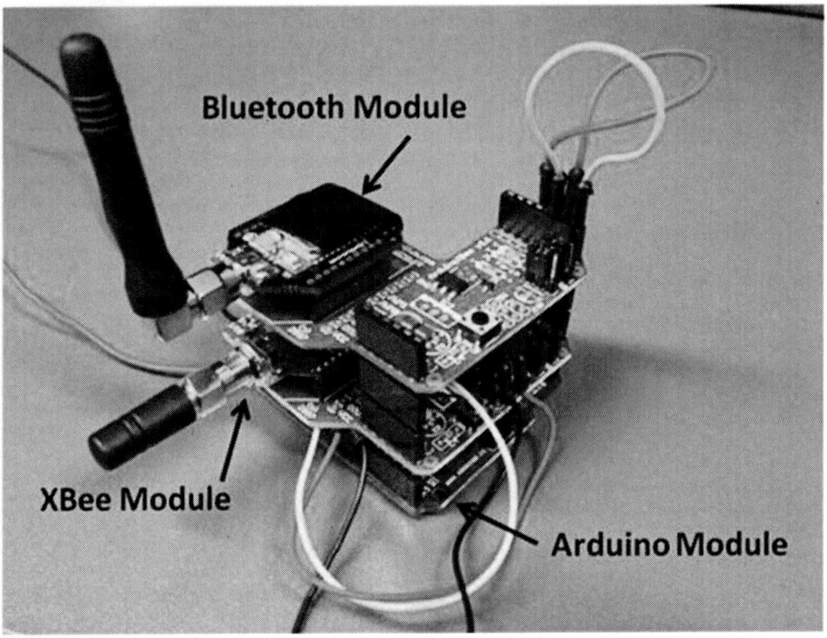

Figure 4. Probe node

During the data collection process at each probe, the information was organized in a consistent manner. Each device frame is 8 bytes and included a marker byte, a 4 byte timestamp, and 3 bytes for the truncated MAC address. To increase privacy, only half of the MAC address is recorded which still provides 16 million unique combinations for one probe network. The first 3 bytes of the packet are reserved bytes; a control byte, a length byte, and a probe id byte. The maximum packet size that can be produced is 100 bytes, as this is the size of the receive buffer of the XBee module.

The XBee module on the master node was set as the network coordinator, and the XBee module on each probe was set to associate with the network coordinator. To save power, the XBee module on the probes were configured to hibernate when not in use. Due to changes in temperature and other internal and external factors on the probes and master node, the calculated time offset at various nodes slowly drifted over time, requiring scheduled clock resynchronization.

The function of the probe is illustrated in the main flow diagram of Fig. 5, and the scan flow sub diagram is illustrated in Fig. 6.

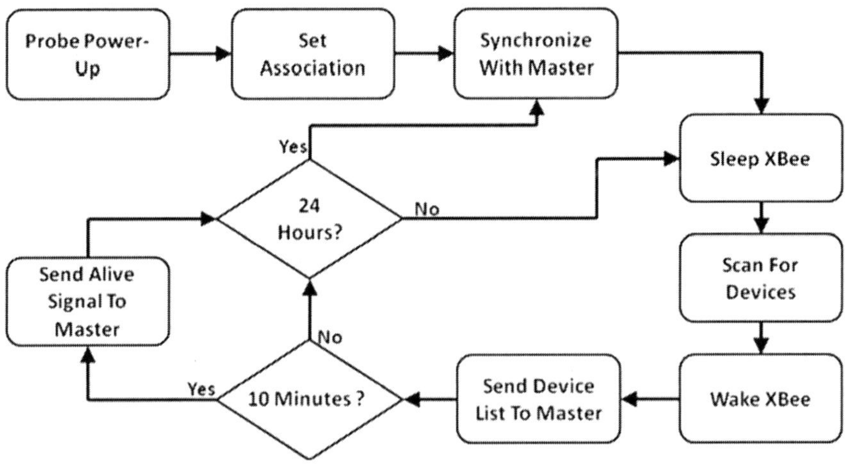

Figure 5. Probe flow diagram

5.3 Master node

The master node consisted of Arduino-based GBoard from iteadstudio.com. For the GSM module, a preexisting library was used, with modifications and additional functionality to meet system requirements.

5.4 Power

The desired minimum run time of each probe was 24 h. A combination of battery-only and battery-plus-solar were used, demonstration the viability of a solar rechargeable source for the wireless sensor network. The rechargeable power sources were two 4,400 mA, 3.7 V lithium ion batteries hooked up in parallel to a step-up converter (to 7 V). Both batteries are connected to lithium-ion chargers powered by a 3.7 W, 6 V solar panel. A custom step-up converter was also designed to have lower current draw compared to the off the shelf step-up converter; thereby increasing the life expectancy of the supply.

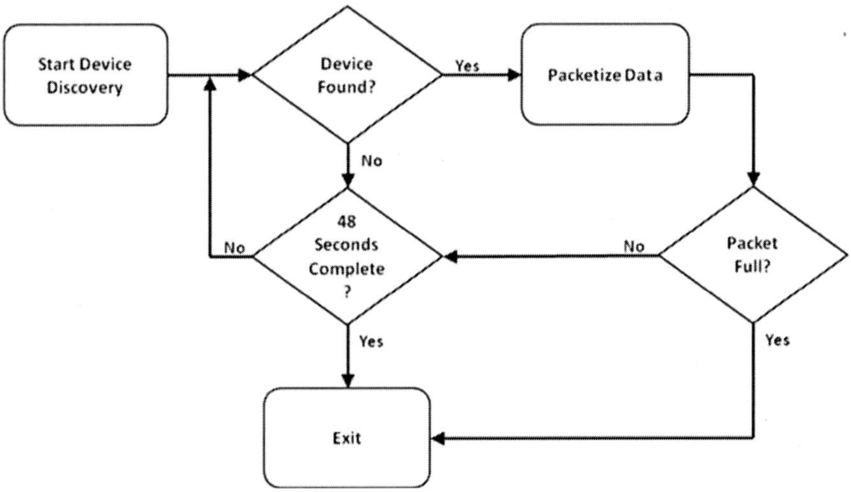

Figure 6. Bluetooth scan diagram

5.5 Database and front-end

A MySQL database was used in order to store data, refreshed at five minute intervals, and display data to a user through a website or a smartphone app. Initial scripts were written where device MACs are compared to a table of existing MACs to find matching devices with a different probe location. When found, a probe-pair was created using the start probe, end probe and detection time difference; this was used to estimate the velocity of the device. Another PHP script was created that automatically runs every five minutes, refreshing the traffic data that is displayed on the websites. The data is stored as the total travel time for all vehicles and the total vehicle count through each probe-pair.

5.6 Implementation

A Bluetooth discovery test was conducted to ensure devices traveling in excess of 80 km/h could be detected. The XBee distance test consisted of testing the connectivity while increasing the separation between master and slave devices. Line of sight difficulties were encountered for distances exceeding 200 m. As a complete test of the system, four probes were placed along a major thoroughfare to capture the traffic over several 24 h timeframes for multiple iterations representative of different traffic and weather conditions. Probes and master nodes were mounted to light standards along the thoroughfare using steel strapping at an elevation of approximately 2.5 m. Several single-probe and multi-probe trials were carried out over winter, 2013. Data were cross-referenced to data obtained by mechanical traffic counters and to cellular service provider data that can serve as proxies for users' movements between cell towers. Qualitatively similar trends in traffic density and flow were observed from the data sources. The reference design was found to capture an average of 4.5 % of real vehicle traffic (when compared to mechanical counters), which is above the 2-3 % conjectured as being required for statistically accurate traffic flower inferencing [6]. Initial findings lent credibility to the reference design as having the characteristics of a viable full-scale system. The cost of the reference design described above was approximately $1,500 in 2012 and 2013.

The reference system described above provides insights into the type of data that can be easily and cost effectively collected at a relatively inexpensive capital cost. The reference design, which was an academic prototype, is validated conceptually by others who report similar systems and investigations. In a study on traffic monitoring with Bluetooth sensors over ZigBee, an intervening multitier network is discussed [30]. Although a different protocol is deployed, the basic ideas and rationales are similar. Others similarly discuss the use of Bluetooth data to infer vehicle proximity as a means of estimating traffic characteristics, including the use of solar power to meet system energy requirements [51, 52].

6. ADDITIONAL CONSIDERATIONS

As of time of writing, the emerging wireless technology for inferring traffic is that based on Bluetooth 2.x. Bluetooth 2.x is the most widely deployed mid-range communication protocol has a range commensurate with distances typical of traffic and traffic control systems. The ubiquity of Bluetooth 2.x and its ready application to traffic-related contexts currently makes it a natural choice for ITS applications. In the future, communication alternatives and superior wireless versions - ostensibly developed for different user applications - will undoubtedly contribute to the data collected for ITS. For example, Bluetooth 4.0 is currently available for 'coin-cell-powered' devices and is currently not obviously applicable to common ITS applications; however, this applicability is likely to develop as new opportunities emerge. The same is true for other technologies, in the spirit that the most effective uses of a technology often emerge after its development rather than as a bounded or fixed a priori specification.

From our experience, one significant recommendation in this field of Bluetooth sensor systems for ITS would be to avoid ad hoc intermediate networks. The difficulties with installation and reliability make data collection networks that rely on lines of sight for communication too constraining. In addition, the XBee communications used in the reference design were power-intensive and not tuned for power conservation. As an alternative, each sensor should ideally be equipped with GSM/GPRS and have the backend service parse data from each probe directly. The additional costs of this approach are likely to be outweighed by the benefits accrued through much simpler deployment and lower maintenance. The intervening XBee, communication network also added complexity through the use of a less efficient protocol for data transmission than GSM directly. This follows the principle of "Lex Parsimonia" where the simplest system is most likely the best.

Future work should investigate the integration of mobile Bluetooth probes as well as stationary probes. This increases processing and analysis of the data collected, but would augment data collected by the stationary Bluetooth probes. An example of a vehicle trajectory while scanning, is in [53], while a patent application for a mobile probe can be found in [54]. Utilizing mobile probes would also allow for augmenting of the data with information such as acceleration at intersections, and would be a direct means of inferring environmental conditions such as ice and snow.

Finally, a decided advantage of a Bluetooth traffic monitoring system through a Services Oriented Architecture is that it can easily and inexpensively augment any existing traffic measurement system.

The real academic, engineering, and organizational challenges will lie in full-scale deployment of a Bluetooth sensor system at many intersections across a large urban centre. The probe sensor network may be augmented by a large number of probe vehicles that could also be configured to upload data from proximate devices detected while in transit. The only requirement for this additional data from probe vehicles is to augment the discovered Bluetooth MAC address and timestamp with GPS information. Once scaled to city-wide deployment, the data mining challenges will be considerable and will require a whole new big data approach, but will also become an invaluable input to an ITS.

ACKNOWLEDGMENTS

The authors thank the City of Winnipeg Public Works Department for financial support, and in particular Mr. Doug Hurl and Mr. Luis Escobar. In addition, the financial support of NSERC and NSERC NCE Auto 21 is gratefully acknowledged.

REFERENCES

1. Research and Innovative Technology Administration (2010). Achieving the Vision: From VII to Intellidrive. [Online]. Available: http://www.its.dot.gov/press/2010/vii2intellidrive.htm

2. R. Sengupta, C. Manasseh and S. Levenson,. "Service oriented architectures for VII – Extending web services to the roadway". [Online]. Available: http://www.its.berkeley.edu/volvocenter/VREF/UCBSOA_VIIChristian.pdf

3. E. Hossain, G. Chow, V.C.M. Leung, R.D. McLeod, J. Mišić, V.W.S. Wong, and O. Yang, "Vehicular telematics over heterogeneous wireless networks: A survey", Computer Communications, Volume 33 (Issue 7, 3), Pages 775–793 (May 2010)

4. J. Sorensen, "One in Evert Crowd", Traffic Technology International, February-March, http://www.dagstuhl.de/Materials/Files/10/10121/10121.XuBo1.Paper.pdf (2010)

5. B.S. Kerner, C. Demir, R.G. Herrtwich, S.L. Klenov, H. Rehborn, M. Aleksic and A. Haug, "Traffic state detection with floating car data in

road networks. Proceedings of the 8th International IEEE Conference on Intelligent Transportation Systems, 44–49 (2005)

6. Herrera, J.C., Work, D.B., Herring, R., Ban, X., Jacobson, Q., Bayen, A.M.: Evaluation of traffic data obtained via GPS-enabled mobile phones: The Mobile Century field experiment". Transportation Research Part C: Emerging Technologies **18**(4), 568–583 (2010)CrossRef

7. G. Halbritter, T. Fleischer, C. Kupsch, J. Kloas and U. Voigt, "Nationale Innovationsstrategien für neue Techniken und Dienste zur Erreichung einer „nachhaltigen Entwicklung" im Verkehr (Report FZKA 7157). Karlsruhe, Germany: Forschungszentrum Karlsruhe GmbH, (2005)

8. Y. Yim and R. Cayford, "Investigation of vehicles as probes using global positioning system and cellular phone tracking: Field operational test", Number UCB-ITS-PWP-2001-9, California PATH Program, Institute of Transportation Studies, University of California, Berkeley, CA, (2001)

9. J.L. Yagnace, C. Drane, Y.B. Yim, and R. de Lacvivier, "Travel time estimation on the San Francisco Bay area network using cellular phones as probes". Number UCB-ITSPWP-2000-18, California PATH Program, Institute of Transportation Studies, University of California, Berkeley, CA, (2000)

10. A.Alessandri, R. Bolla, and M. Repetto, "Estimation of freeway traffic variables using information from mobile phones", In Proc. American Control Conference the 2003, volume 5, pages 4089–4094, Denver, CO, (June 2003)

11. R. Neighbour, R.D. McLeod, M.R. Friesen, and S.N. Mukhi, "Vehicular Traffic Modeling Governed by Cellular Phone Trajectories", VTC-Fall (2012)

12. BlueTags to install world's first Bluetooth tracking system, 2003. http://www.geekzone.co.nz/content.asp?contentid=1070.

13. M. Haase and M. Handy. BlueTrack – Imperceptible tracking of bluetooth devices. In Ubicomp Poster Proceedings, (2004)

14. Y. Wen, J.L. Pan and J.F.Le, "Survey On Application Of Wireless Sensor Networks For Traffic Monitoring", International Conference on Transportation Engineering 2007 (ICTE 2007)

15. E. Minge, J. Kotzenmacher, and S. Peterson "Evaluation of Non-Intrusive Technologies for Traffic Detection", Mn. DoT., Final Report #2010-36, http://www.miovision.com/wp-content/uploads/ Evaluation_of_Non-Intrusive_Traffic_Detection_Technologies_pdf (September2010)

16. Barceló, J., Montero, L., Marqués, L., Carmona, C.: Travel Time Forecasting and Dynamic Origin–destination Estimation for Freeways Based on Bluetooth Traffic Monitoring". J. Transp. Res. Board **2175**, 19–27 (2010)CrossRef

17. Y. Wang, Y. Malinovskiy, Y-J. Wu and U-K. Lee, "Field Experiments with Bluetooth Sensors", n.d., http://www.westernstatesforum.org/ Documents/UW_FINAL_Bluetooth_WSRTF_6-17-10.pdf (2010)

18. Wasson, J.S., Sturdevant, J.R., Bullock, D.M.: Real-Time Travel Time Estimates Using MAC Address Matching. Institute of Transportation Engineers Journal, ITE **78**(6), 20–23 (2008)

19. R. Haseman, J.S.Wasson and D.M. Bullock, D M., Real-Time Measurement of Travel Time Delay in Work Zones and Evaluation Metrics Using Bluetooth Probe Tracking. Transportation Research Record: Journal of the Transportation Research Board, pp. 40–53, (2010)

20. M. Asudeqi, "Optimal Number and Location of Bluetooth Sensors for Travel Time Data Collection in Networks", M.Sc. Thesis, University of Maryland, http://hdl.handle.net/1903/9400

21. H. Sawant, J. Tan, Q. Yang, and Q. Wang, "Using Bluetooth and sensor networks for intelligent transportation systems," Intelligent Transportation Systems, 2004. Proceedings. The 7th International IEEE Conference on, vol., no., pp.767,772, 3–6, doi: 10.1109/ITSC.2004. 1398999 (Oct. 2004)

22. B. Araghi, K.S. Pederson, L.T. Christenson, R. Krishnan, and H. Lahrmann, "Accuracy of Travel Time Estimation using Bluetooth Technology: Case Study Limfjord Tunnel Aalborg", ITS World Congress, (2012)

23. Nielson, "Smartphone Switch: Three Fourths of Recent Acquirers Chose Smartphones" Available: http://www.nielsen.com/us/en/newswire/

2013/smartphone-switch--three-fourths-of-recent-acquirers-chose-smart.html

24. Barceló, J., Montero, L., Marqués, L., Carmon, C.: Travel Time Forecasting and Dynamic Origin–destination Estimation for Freeways Based on Bluetooth Traffic Monitoring". J. Transp. Res. Board **2175**, 19–27 (2010)CrossRef

25. J.D. Porter, D.S. Kim and M.E. Magana, "Wireless Data Collection System for Real-Time Arterial Travel Time Estimates", OTREC 10–16, http://www.oregon.gov/ODOT/TD/TP_RES/docs/reports/2011/wirelessdata.pdf, (March 2011)

26. S.E. Young, "Bluetooth Traffic Detectors for Use as Permanently Installed Travel Time Instruments", STMD -MDOT/SHA, http://roads.maryland.gov/OPR_Research/MD-12-SP909B4D-Bluetooth-Traffic-Detectors_Report.pdf, (Feb. 2012)

27. Anonymous Wireless Address Matching (AWAM) proof of concept demonstration, Texas Transportation Institute, http://ttihouston.tamu.edu/bluetooth/projects.aspx

28. Porter, J.D., Kim, D.S., Magaña, M.E., Poocharoen, P., Arriaga, C.A.G.: Antenna Characterization for Bluetooth-Based Travel Time Data Collection". Journal of Intelligent Transportation Systems: Technology, Planning, and Operations **17**(2), 142–151 (2013). doi:10.1080/15472450.2012.696452CrossRef

29. J.D. Porter, D.S. Kim, M.E. Magaña, P. Poocharoen and C.A.G. Arriaga, "Antenna Characterization for Bluetooth-Based Travel Time Data Collection", http://www.westernstatesforum.net/Documents/2011/presentations/OSU_KimEtAl_Final_AntennaCharacterization.pdf.

30. Vehicle Traffic Monitoring Platform with Bluetooth Sensors over ZigBee, [Online]. Available: http://sensor-networks.org/index.php?page=1129929946

31. Klunder, Y. S. Chen, K. Zantema, and K. Friso. "Consistent Traffic Modeling and Calibration at Different Resolution Levels Based on Various Sensor Data for Real-time Traffic Management", 2nd International Conference on Models and Technologies for Intelligent Transportation Systems, 22–24, Leuven, Belgium (June, 2011)

32. Bachmann, C., Roorda, M.J., Abdulhai, B., Moshiri, B.: Fusing a Bluetooth Traffic Monitoring System With Loop Detector Data for Improved Freeway Traffic Speed Estimation. Journal of Intelligent Transportation Systems: Technology, Planning, and Operations **17**(2), 152–164 (2013).

33. E. Hidle, "Bluetooth can help local traffic flow", http://www.columbian.com/news/2013/oct/21/bluetooth-traffic-flow-clark-county/

34. S. Yucel, H. Tuydes-Yaman, O. Altintasi, M. Ozen, "Determination of Vehicular Travel Patterns in an Urban Location using Bluetooth Technology", https://itswc.confex.com/itswc/AM2013/webprogram/ExtendedAbstract/Paper11460/ITS_America_sule_yucel.pdf

35. M. Blogg, C. Semler, M. Hingorani, R. Troutbeck, "Travel Time And Origin–destination Data Collection Using Bluetooth MAC Address Readers", Australasian Transport Research Forum 2010 Proceedings, p. 1–15 (2010)

36. D.D. Puckett and M.J. Vickich, "Bluetooth-Based Travel Time/Speed Measuring Systems Development", Texas Transportation Institute, DTRT06-G-0044, http://utcm.tamu.edu/publications/final_reports/Puckett_09-00-17.pdf, (2010)

37. J. Effinger, A. J. Horowitz, Y. Liu, and J. Shaw, "Bluetooth Vehicle Reidentification for Analysis of Work Zone Diversion." In Transportation Research Board 92nd Annual Meeting, no. 13–2159. (2013)

38. S. Zinner, "A methodology for using bluetooth to measure real-time work zone travel time." PhD diss., Georgia Institute of Technology, 2012

39. C.M. Day, J. S. Wasson, T. M. Brennan, and D. M. Bullock. Application of Travel Time Information for Traffic Management. Publication FHWA/IN/JTRP-2012/06. Joint Transportation Research Program, Indiana Department of Transportation and Purdue University, Indiana

40. Steiner and B. Sick, "A new method for travel time estimation on long freeway sections", Institute of Data Analysis and Process Design, Zurich University of Applied Sciences. European Journal of Transport and Infrastructure Research (ISSN 1567–7141), Vol.8, Issue 4, (2008)

41. S. Hay and R. Harle, "Bluetooth tracking without discoverability," Proc. of the 4th Int'l Symp. on Location and Context Awareness. Springer-Verlag, pp. 120–137 (2009)

42. H. Quintero, L. Felipe, F.M. Pérez, D.M. Jorquera, and Virgilio Gilart Iglesias. "SOA-based Model for the IT Integration into the Intelligent Transportation Systems." IEEE ITSC2010 Workshop on Emergent Cooperative Technologies in Intelligent Transportation Systems, (2010)

43. X. Tao, C. Jiang, Y. Han, "Applying SOA to intelligent transportation system", Proceedings of the IEEE International Conference on Services Computing (SCC'05), Vol.2, pp. 101–106 (2005)

44. G. Chakraborty, K. Naik, D. Chakraborty, N. Shiratori and D. Wei, "Analysis of the Bluetooth device discovery protocol. Wireless Networks", Wireless Networks

45. Saeedi, S. Park, D.S. Kim, J.D. Porter, "Improving Accuracy and Precision of Travel Time Samples Collected at Signalized Arterial Roads with Bluetooth Sensors ', Volume 2380, pp. 90–98, (2013)

46. F. Zheng, "Modelling Urban Travel Times", Thesis Delft University of Technology, http://repository.tudelft654/Dissertation_Zheng.pdf (2011)

47. J.D. Lees-Miller, R.E. Wilson, and S. Box. "Hidden Markov Models for Vehicle Tracking with Bluetooth." In Transportation Research Board 92nd Annual Meeting, no. 13–3032. (2013)

48. M. Friesen, R. Jacob, P. Grestoni, T. Mailey and R.D. McLeod, "Vehicular Traffic Monitoring using Bluetooth", CCECE 2013

49. R. Cavaliere, P. Valleri, R. Kloibhofer, "The INTEGREEN approach for the assessment of environmental traffic management policies, 9th ITS European Congress, Dublin, Ireland, 4–7 June 2013

50. F. Biora, F. d'Aprile, and R. Marino, "A Large Scale Application for Bluetooth-based Travel Time Measurement in the Netherlands", 19th ITS World Congress, Vienna, Austria, 22 to 26 October (2012)

51. J. Benavides, B.C.P. Demianyk, S.N. Mukhi, M. Laskowski, M.R. Friesen and R.D. McLeod, "Smartphone Technologies for Social Network Data

Generation and Infectious Disease Modeling", J. of Medical and Biological Engineering, 32(4): 235–244

52. L.M. Peterson, "Obtaining vehicle traffic information using mobile Bluetooth detectors", https://www.google.com/patents/US20120276847, Google Patents (2012)

Chapter 6

Detection of Emergency Telephone Indicators in a Tunnel Environment

Zhipeng Wang[1], Matasaka Kagesawa[1], Shintaro Ono[1], Atsuhiko Banno[2], Takeshi Oishi[1], Katsushi Ikeuchi[1]

[1]Institute of Industrial Science, The University of Tokyo, Tokyo, Japan

[2]Advanced Industrial Science and Technology, Ibaraki, Japan

ABSTRACT

Positioning of vehicles is important for ITS. In a tunnel environment, most positioning solutions based on GPS sensors or ordinary cameras will fail. For positioning we propose a method to detect emergency telephone indicators in a tunnel environment by using infrared cameras. The proposed detection method makes use of both appearance and motion information of the target objects. By optimizing the detecting pipeline, the method works in real time and produced 100 % detection rate and 0 % false alarm rate in one of our experiments.

KEYWORDS

Object detection, Positioning,

1. INTRODUCTION

Positioning of vehicles acts as a fundamental role in autonomous driving, and is also of great importance for driving assistance, vehicle navigation, etc. When GPS sensors function properly, the task is easy. While in a tunnel environment, there are no GPS signals available for most of the time. A new positioning system which functions properly in a tunnel environment is necessary [6]. In this paper we propose an object detection method which can be used for positioning systems in tunnels.

Our proposal is part of an automated driving system in a NEDO project, "Development of Energy-saving ITS Technologies". The automated driving system is vehicle-oriented, and an express way is the main application scenario. No specific facilities are assumed to exist on road sides, while instead, the experimental vehicles are equipped with sensors and vehicle-to-vehicle communication systems. There are a few sensors used for positioning. For example, sensors used to extract white road lines to estimate the lateral position in the lane, GPS sensors, dead reckoning systems, and stereo far-infrared camera systems intended for obstacle detection. On the street, GPS sensors can be used for positioning. While positioning in tunnels is difficult since GPS signals are not available and no specific equipment on road sides is assumed. For positioning in tunnels, GPS sensors are used to record the position of tunnel entrances, and dead reckoning systems are used to infer position by continuously sensing the vehicles' speed and direction. However, errors of dead reckoning systems will accumulate. Thus, the proposed method uses far-infrared cameras installed on the vehicles to detect signs in the tunnels, which contain position information, and will be used to eliminate the accumulated errors.

In most tunnels on the expressways in Japan, there are many signs appearing at equal intervals. We focus on the emergency telephone indicators, which appear every 200 meters in tunnels. The absolute coordinates of the emergency telephone indicators can be obtained by the method of [30]. If the emergency telephone indicators can be sensed while traveling in tunnels, and the distance from the vehicle to the indicators can be estimated, then this information can be used to eliminate accumulated errors of dead reckoning systems. Detection methods, e.g., [29], based on ordinary cameras fail due to darkness. Here we use far-infrared cameras, which are suitable in dark environments and are already installed on our experimental vehicles. Inspired by a previous work [28], we propose an approach to detect emergency telephone indicators.

Detection performance and efficiency are the two important aspects of our method. In a tunnel environment, in addition to the target objects, a lot of noisy objects also appear, e.g. ordinary lights, other vehicles, and other vehicles' shadows. And some of the noisy objects cannot even be distinguished from the target objects by appearance, as shown in Fig. 1. The clutter property of the sensed data makes the detection challenging. Our method meets this challenge by making use of both appearance and temporal information of the target objects. There are two main steps in the method. The first step deals with keypoints. It takes original data as input, and outputs keypoint clusters as detection hypotheses. In this step, keypoints are detected, verified and then clustered. To detect keypoints, all points on each frame are uniformly sampled and filtered with pre-set intensity thresholds. Then the keypoints are verified by a simple keypoint appearance model built by *k*-means. At the end of the first step, the keypoints are clustered based on the Euclidean distance. The second step takes the keypoint clusters as input, verifies them by appearance and temporal information, and outputs the keypoints which pass verification, as detection results. In the second step, the keypoint clusters are labeled based on appearance by an Adaboost machine, which is trained using intensity histograms of keypoint clusters from target objects and keypoint clusters from noisy objects. The keypoint clusters are also tracked by temporal association through frames. Motion information encoded in the trajectories are used to further verify the keypoint clusters. Finally, the keypoint clusters which pass both appearance and temporal verifications are confirmed as emergency telephone indicators.

This pipeline is also designed with consideration for the requirement of efficiency. The method deals with the large amounts of information contained on one frame, following a hierarchical manner. The later a step is, the more time-consuming it is, and the fewer instances it deals with. From an image containing 10^5 pixels, 10^4points go through the keypoint detection step of testing by intensity thresholds. Then in average, 10^3 keypoints are detected, and verified, leaving about 10^2 keypoints to be clustered. Afterwards, fewer than 10 keypoint clusters are left; these are dealt with by the very time-consuming steps of generating image features and tracking.

Figure 1. Original data and detection results. In (**a**), the *red arrow* points to the target object: emergency telephone indicator, and the *green arrows* point to noisy objects. In (**b**), *red rectangles* mark detection hypotheses labeled as positive using appearance information, and *green rectangles* mark negative ones. *Yellow*trajectories mark detection hypotheses labeled as positive using temporal information, and *white* trajectories mark negative ones

The advantage of our method is its ability to give promising detection results from cluttered data in real time. In addition, this method successfully combines bottom-up and classification methods, as well as combines both appearance and temporal information.

The paper is organized as follows: Section 2 reviews related work, Section 3proposes a pipeline for the method, Section 4 gives experimental results, and Section 5 concludes.

2. RELATED WORK

Most modern detection methods fall into two categories. Some [5, 8, 10, 14, 19, 25,27, 32] follow the sliding-window schema, and they detect objects by considering whether each of the sub-images contains an instance of the target object. Classifiers are usually employed by these methods. The other methods [1, 7, 9, 16,20–22] infer object centers based on local image features in a bottom-up manner. The proposed method takes advantages of both frameworks. Following the bottom-up manner, keypoints are detected, verified, and clustered. After these steps, the keypoint clusters are considered as detection hypotheses. Then following the sliding-window schema, the keypoint clusters are verified by their appearance and temporal information, using discriminative methods. Previous methods [24] also consider the combination of the two frameworks. Detection hypotheses are gained using a Hough transform and then verified by support vector machines in [20, 31]. The methods in [11, 23], use randomized decision trees for both decisions: whether local features belong to foreground objects or not, and decisions of their Hough votes. The method proposed in [15] describes both frameworks in the same manner. While giving state-of-the-art detection performance, these other methods can't meet the requirement for efficiency as our method does. Our work is also related to feature grouping methods [31], detecting methods using trajectories [3, 4], tracking methods [12, 17], and methods integrating appearance and temporal information [29]. Especially, compared with the method proposed in [28], our method employs a more effective classifying machine by setting biased weights for positive and negative training examples, and far over-performs [28].

3. EMERGENCY TELEPHONE INDICATOR DETECTION

Our proposal can be considered a two-step method. The first step deals with keypoints. It takes original data as input, and outputs keypoint clusters as detection hypotheses. The second step takes these keypoint clusters as input, verifies them by their appearance and motion information, and outputs the ones which pass verifications as detection results.

3.1 Keypoint Detection

In data collected using ordinary cameras, keypoints [2, 18] invariant to rotations, affine changes, and illumination changes are preferable. In our case, keypoint detection is intended to provide hypotheses for emergency telephone indicators. Thus intensity is of great importance. Our method employs a simple yet useful method to detect keypoints. Firstly, points are uniformly sampled for an offset of 6 in width, and 7 in height (the length of an emergency telephone indicator is larger than its width). In this manner the magnitude of instances is reduced by nearly two orders. Then points that pass the test, which verifies them by setting intensity thresholds, are considered as keypoints. Here a Gaussian distribution is assumed for the intensities of the points.

let {x} denote all the sampled points, I_x the intensity of each point, and l_x the label. If the point is considered as belonging to emergency telephone indicators, l_x= 1, otherwise, l_x= 0. By setting lower threshold, Ith1xIxth1, and higher threshold, Ith2xIxth2, the probability that points belongs to emergency telephone indicators based on their falling into this interval is given by,

$$P\left(l_{\mathbf{x}} = 1 \mid I_{\mathbf{x}}^{th1} \leq I_x \leq I_{\mathbf{x}}^{th2}\right) = \frac{P\left(l_{\mathbf{x}} = 1, I_{\mathbf{x}}^{th1} \leq I_x \leq I_{\mathbf{x}}^{th2}\right)}{P\left(I_{\mathbf{x}}^{th1} \leq I_x \leq I_{\mathbf{x}}^{th2}\right)} \quad (1)$$

At this step, the possibility that as few points as possible, belonging to the emergency telephone indicators, are excluded, is also considered. The probability of one point falling into the defined interval based on its belonging to emergency telephone indicators is given by,

$$P\left(I_{\mathbf{x}}^{th1} \leq I_x \leq I_{\mathbf{x}}^{th2} \mid l_{\mathbf{x}} = 1\right) = \frac{P\left(l_{\mathbf{x}} = 1, I_{\mathbf{x}}^{th1} \leq I_x \leq I_{\mathbf{x}}^{th2}\right)}{P\left(l_{\mathbf{x}} = 1\right)} \quad (2)$$

Points for which the intensities fall in the pre-set thresholds, are detected as keypoints.

3.2 Keypoint Verification

As shown in Fig. 2b, the detected keypoints don't just belong to emergency telephone indicators, but also belong to the background. For training purposes, keypoints belonging to emergency telephone indicators are considered positive, all others are negative.

To verify the keypoints, the appearance of the sub-image around each keypoint is used. Intensity histograms are used to describe the appearance. Noisy keypoints not only come from the wall of the tunnel, but also from ordinary lights, other vehicles, and other vehicles' shadows. Thus robust linear classifiers are not suitable for the verification. Here, a general model in the form of a simple mixture is used. The *k*-means method is used to cluster the intensity histograms, $\{A_{\mathbf{x}}, l_{\mathbf{x}} = 1\}$ of the positive keypoints, and, $\{A_{\mathbf{x}}, l_{\mathbf{x}} = 0\}$ of the negative keypoints.

a b

Figure 2. Keypoint detection

Let $\{C_1^i, i = 1, 2, \ldots, n_1\}$ denote the intensity histogram centers of the positive keypoints, and $\{C_0^i, i = 1, 2, \ldots, n_2\}$ the negative. For each C_1, the average Euclidean distance between $\{C_0^i, i = 1, 2, \ldots, n_2\}$ is calculated as,

$$Eu\left(C_1^i\right) = \frac{1}{n_2} \sum_{j=1}^{n_2} Euclid\left(C_1^i, C_0^j\right) \tag{3}$$

Here, $Euclid$ calculates the Euclidean distance, and Eu is an evaluation function of the positive feature centers. The positive feature centers are ranked by Eu and the 10 positive feature centers with the largest Eu are chosen and used for verification.

For verification, the intensity histogram of each keypoint's surrounding sub-image is extracted. Then the Euclidean distance between the extracted intensity histogram and its nearest positive feature center is calculated. If this distance exceeds a threshold, $D^{th}{}_{Ax}$, it is considered negative, otherwise it is considered positive. Here, for simplicity, unlike [26], the same threshold is used for all components of the mixture.

3.3 Keypoint Clustering

After the keypoint verification step, on some frames the result is pretty good, while on other frames appearance of the keypoints is not enough to decide whether the keypoints belong to the emergency telephone indicators or not. Here generation of keypoint trajectories is not feasible, since nearby keypoints are similar in appearance and the time complexity of associating such a large number of keypoints along the time dimension is high. So the keypoints are clustered, then data association in time dimension only is needed to deal with a small number of keypoint clusters.

To cluster the keypoints, a minimum spanning tree (mst) is built using the pairwise Euclidean distance between two keypoints. Then the mst is split by cutting edges larger than a threshold. This results in a grouping of the keypoints, denoted by, $\gamma = \{g\}$.

3.4 Keypoint Cluster Verification by Appearance

For each keypoint cluster, the smallest bounding rectangle is considered a detection hypothesis, as shown in Figs. 3c and 3d. There are three main sources of noise: ordinary lights, other vehicles, and other vehicles' shadows. The global appearance of ordinary lights is different from that of the emergency telephone indicators'. As ordinary lights get further from the infrared camera, the intensities of their corresponding sub-images in the collected data gets lower. At a certain distance, the intensities of the ordinary lights are almost the same as the intensities of the emergency telephone indicators'. For ordinary lights of which the intensities are higher than the intensities of emergency telephone indicators', the transition regions from the lights to tunnel walls will have similar intensities as the emergency telephone indicators'. This indicates that although locally the emergency telephone indicators share the same appearance with ordinary lights, globally they can still be distinguished by appearance. As for other vehicles and their shadows, their intensity range is very close to the intensity range of the emergency telephone indicators', and they can hardly be distinguished by appearance alone.

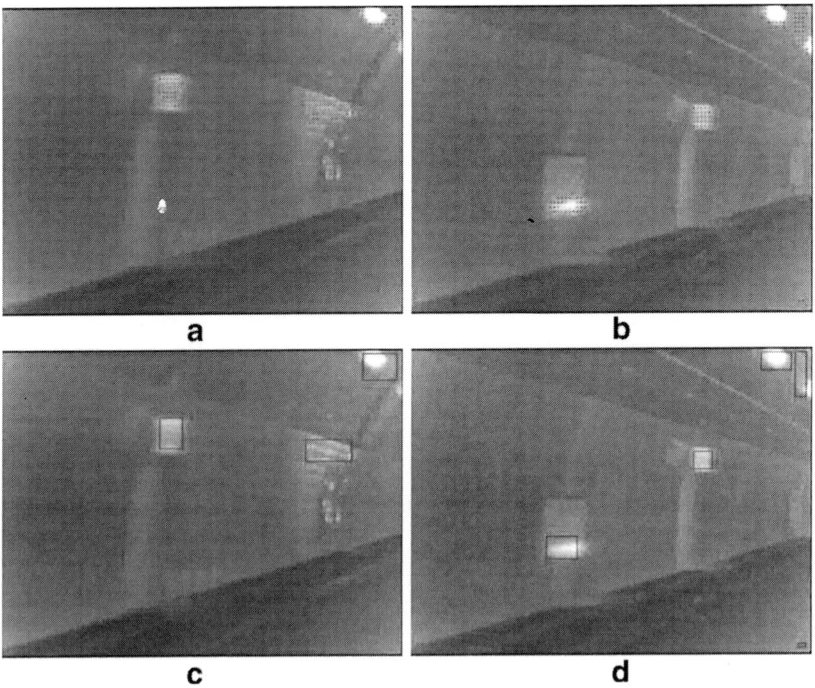

Figure 3. Keypoint verification and clustering. *Red circles* mark keypoints which pass the verification, while *blue marks* failed ones. *Rectangles mark* keypoint clustering results

At this step, the keypoint clusters are verified by their appearance, ideally excluding keypoint clusters belonging to ordinary lights. An Adaboost machine is trained using intensity histograms of the emergency telephone indicators and ordinary lights. The appearance of other vehicles is close to that of the emergency telephone indicators, and they are not used for training the machine. For training of the machine, labeled 32-dimensional intensity histograms are firstly normalized. Then each weak classifier of the machine makes a decision on one dimension of the intensity histograms. After this step, each keypoint cluster is either labeled as positive or negative.

In this step, to emphasize the Adaboost machine's performance on the positive training examples, we set the initial weights of the positive training examples 7 times as large as the weights of the negative training examples. Since in practice, whether each keypoint cluster is a target object or not is decided by both appearance and motion information. The difficulties of excluding noisy objects can be left for later steps.

3.5 Keypoint Cluster Tracking

Not all noisy detection hypotheses can be excluded by using appearance, as shown in Fig. 4. To distinguish keypoint clusters belonging to other vehicles and their shadows, the keypoint clusters are tracked through frames to generate trajectories.

Figure 4. Keypoint cluster verification by appearance. *Red rectangles*: positive detection hypotheses, and *green*: negative detection hypotheses

In our case of keypoint cluster tracking, the problem is relatively simple, since no occlusion occurs. To keep the method on-line and maintain efficiency, a pool of trajectories are kept, $\tau = \{T_g^i, i = 1, 2, \ldots, n\}$ and new detection hypotheses act as detection responses, v= $\{n_g^i, i = 1, 2, \ldots, m\}$, in tracking. The problem of tracking is modeled by finding the best data association hypothesis, H^*, between the trajectory set and detection response set as,

$$\begin{aligned} H^* &= \arg\max_{H \in \eta} P(H, \nu) \\ &= \arg\max_{H \in \eta} \left(\prod_{\left(T_g^i, n_g^j\right) \in H} P_{link}\left(n_g^j, T_g^i\right) \right) \end{aligned} \tag{4}$$

Let u_{ij} = 1 or 0 indicate if njgngj is linked to TigTgi or not, and assuming each trajectory can link once, and each detection response can only be linked once, the problem can be modeled as,

$$\arg\max_{u_{ij}} \sum_{i=1}^{n} \sum_{j=1}^{m} u_{ij} \ln P_{link}\left(n_g^j, T_g^i\right)$$

$$s.t.: \ u_{ij} = 0 \text{ or } u_{ij} = 1, \forall i, \forall j;$$

$$\sum_{i=1}^{n} u_{ij} \leq 1 \ ; \sum_{j=1}^{m} u_{ij} \leq 1 \ ($$

Here, $P_{link}\left(n_g^j, T_g^i\right)$ is defined by the appearance difference, the scale difference, and the time gap between the last detection response contained in TigTgi and njgngj. While the Hungarian algorithm [13] gives a near-optimal solution, we follow a very simple manner for the solution by finding the best matched pairs and excluding them until no matching pairs can be found.

3.6 Keypoint Cluster Verification by Motion

As shown in Fig. 5, the trajectories from keypoint clusters belonging to emergency telephone indicators are different from other objects' trajectories. In this step, the temporal information encoded in the trajectories is used to further verify the keypoint clusters. A linear model is used to fit each trajectory, and the Pearson Correlation Coefficient (PCC) of the fitting is the criteria for the decision. Let $\left(x_g^i, y_g^i\right)$ denote the coordinate of the *i*th element belonging to a trajectory. The linear assumption is that $y_g^i = a_0 + a_1 x_g^i$. The PCC of the fitting is defined as,

$$r = \left| \frac{\sum_i \left(x_g^i - \bar{x}_g\right)\left(y_g^i - \bar{y}_g\right)}{\left[\sum_i \left(x_g^i - \bar{x}_g\right)^2 \cdot \sum_i \left(y_g^i - \bar{y}_g\right)^2\right]^{1/2}} \right| \quad (5)$$

Where r is used to decide if the trajectories of the keypoint clusters belong to emergency telephone indicators or not.

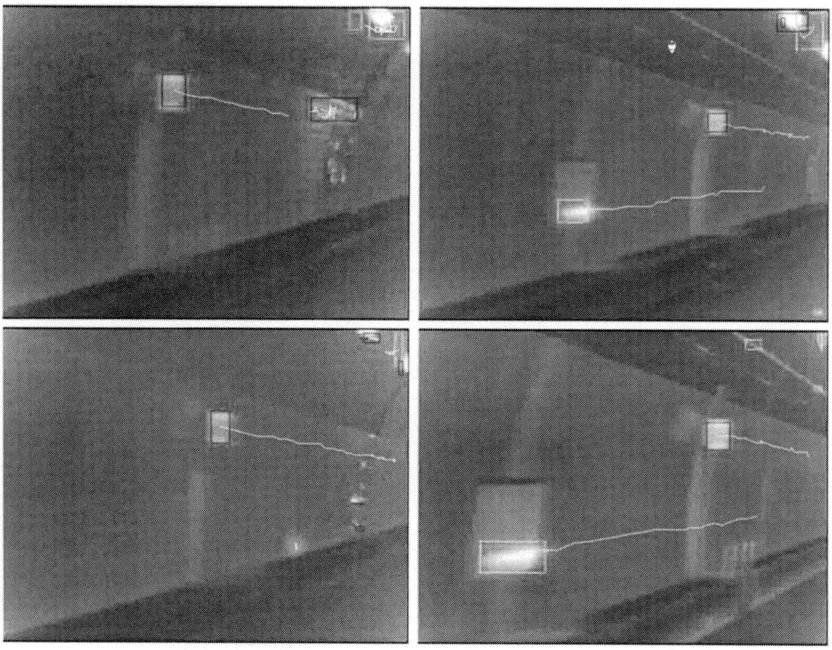

Figure 5. Detection results

3.7 Object Detection

For each keypoint cluster on the current frame, there exists a label given by the Adaboost machine according to its appearance, and the likelihood of fitting its trajectory to a straight line. For each keypoint cluster, it is considered an emergency telephone indicator if and only if its label - given by the Adaboost machine - is positive, its trajectory is longer than l^{th}, and the likelihood of fitting its trajectory to a straight line is larger than r^{th}.

Each trajectory not only connects the detection responses, but also connects the decisions for detection responses made by their appearance and motion patterns. The target objects and noisy objects actually appear in successive frames, and even if we make a wrong decision on one frame, we can expect to recover from this mistake based on the results of other frames. The final results are based on the trajectories of decisions. When one trajectory ends, if more than 80 % of the decisions it connects are positive, then this trajectory is considered positive.

4. EXPERIMENTAL RESULTS

We evaluate our method based on detection performance and efficiency.

4.1 Data

To collect data, we mount an infrared camera on top of the experimental vehicle, and then take several tours of the Awagatake tunnel. Approximately 7,000 frames are collected for each tour. The frame size is 640 × 480, the intensity range is [0,255], and the frame rate is 30 frames per second.

4.2 Implementation Settings

All models are trained using data from the same tour, while evaluated using data from another tour.

To set intensity thresholds for keypoint detection, Gaussian distribution is assumed for the points belonging to emergency telephone indicators. Following the 3σprinciple, Ith1xlxth1 is set to 160 and Ith2xlxth2 to 190. The approximate sub-images of the emergency telephone indicators are manually marked, and used for training the mixture model of keypoint verification. The detected keypoints falling into the sub-image are marked as positive, all other points are marked negative. Note that this model and the training method may not be very accurate, since more accurate marking requires more manual effort. About 30,000 intensity histograms of the positive keypoints are sampled, and about 3,000,000 of the negative. When using of*k*-means for clustering the positive intensity histograms, *k* is set to 40; it is set to 400 for the negative. By using these *k* values, both feature sets are over-segmented. The threshold to verify keypoints DthAxDAxth is set to 0.14 for the normalized histograms. For keypoint clustering, the threshold to split the mst is set to 40, which is half the largest height of the emergency telephone indicators. The Adaboost machine used to distinguish other vehicles and their shadows is trained by intensity histograms of positive keypoint clusters and negative keypoint clusters. We manually mark positive and negative keypoint clusters. If the Adaboost machine is trained by averagely weighted training examples, its correct rate on the training examples is overall 84 %. When trained using our bias weighted training examples, its correct rate is 94 % for the positive training examples, and 77 % for the negative training examples. During keypoint cluster tracking, whether a detection response can be linked to a trajectory or not is constrained by position and scale changes. Here scale change limit is set to 4. When the trajectories are fitted as lines, the linear model is also used in associating new detection responses.

4.3 Detection Results

Using an ordinary desktop computer with Intel Core2 Quad 2.6GHz processors, the method deals with real data at a frame rate of 41 frames per second, and this fulfills real-time requirements.

The detection rate and false alarm rate are evaluated on the keypoint clusters as shown in Table 1. More detection results are shown in Fig. 5.

While evaluated on a much smaller dataset, the detection rate and false alarm rate of [28] are 90 % and 19 % respectively. Our experiment outperforms [28], because our sensed images are much clearer, and also because of our more effective training of the Adaboost machine.

The results on the trajectories of decisions are also evaluated. The method correctly detects all of the 22 emergency telephone indicators with no false alarms. The detection rate is 100 %, and the false alarm rate is 0 %.

Table 1. Detection rate and false alarm rate

Total number	472
Correctly labeled	468
Miss detections	4
False alarms	22
Detection rate	99.2 %
False alarm rate	0.7 %

5. CONCLUSION

We propose an object detection method to detect emergency telephone indicators in a tunnel environment. The method makes use of appearance and motion information of the target objects in a hierarchical manner. With careful optimization of the detection pipeline, the method gives promising results in real time. Based on the detection results, a positioning system in a tunnel environment can be expected.

ACKNOWLEDGMENTS

The work is in part supported by NEDO. The first author is sponsored by China Scholarship Council. The authors thank Patricia Knapp and Kent Fujiwara for revising suggestions on the manuscript.

REFERENCES

1. Barinova, O., Lempitsky, V., Kohli, P.: On detection of multiple object instances using hough transforms. In: CVPR, pp. 2233–2240 (2010)
2. Bay, H., Tuytelaars, T., Van Gool, L.: Surf: speeded up robust features. In: ECCV, pp. 404–417 (2006)

3. Brostow, G., Cipolla, R.: Unsupervised bayesian detection of independent motion in crowds. In: CVPR, vol. I, pp. 594–601 (2006)

4. Brox, T., Malik, J.: Object segmentation by long term analysis of point trajectories. In: ECCV, vol. V, pp. 282–295 (2010)

5. Dalal, N., Triggs, B.: Histograms of oriented gradients for human detection. In: Schmid, C., Soatto, S., Tomasi, C. (eds.) CVPR, pp. 886–893 (2005)

6. Davidson, P., Hautamaki, J., Collin, J.: Using low-cost mems 3d accelerometer and one gyro to assist gps based car navigation system. In: International Conference on Integrated Navigation Systems (2008)

7. Felzenszwalb, P., Huttenlocher, D.: Pictorial structures for object recognition. IJCV **61**(1), 55–79 (2005)

8. Felzenszwalb, P.F., McAllester, D.A., Ramanan, D.: A discriminatively trained, multiscale, deformable part model. In: CVPR (2008)

9. Fergus, R., Perona, P., Zisserman, A.: Object class recognition by unsupervised scale-invariant learning. In: CVPR, vol. II, pp. 264–271 (2003)

10. Ferrari, V., Jurie, F., Schmid, C.: Accurate object detection with deformable shape models learnt from images. In: CVPR (2007)

11. Gall, J., Lempitsky, V.: Class-specific hough forests for object detection. In: CVPR, pp. 1022–1029 (2009)

12. Huang, C., Wu, B., Nevatia, R.: Robust object tracking by hierarchical association of detection responses. In: ECCV, vol. II, pp. 788–801 (2008)

13. Kuhn, H.W.: The hungarian method for the assignment problem. Nav. Res. Logist. Q. 83–97 (1955)

14. Lampert, C., Blaschko, M., Hofmann, T.: Beyond sliding windows: object localization by efficient subwindow search. In: CVPR, pp. 1–8 (2008)

15. Lehmann, A., Leibe, B., Van Gool, L.: Fast prism: branch and bound hough transform for object class detection. IJCV **94**(2), 175–197 (2011)

16. Leibe, B., Schiele, B.: Interleaved object categorization and segmentation. In: BMVC, pp. 759–768 (2003)

17. Leibe, B., Schindler, K., Van Gool, L.: Coupled detection and trajectory estimation for multi-object tracking. In: ICCV, pp. 1–8 (2007)

18. Lowe, D.G.: Distinctive image features from scale-invariant keypoints. IJCV **60**, 91–110 (2004)

19. Maji, S., Berg, A.C., Malik, J.: Classification using intersection kernel support vector machines is efficient. In: CVPR (2008)

20. Maji, S., Malik, J.: Object detection using a max-margin hough transform. In: CVPR, pp. 1038–1045 (2009)

21. Mikolajczyk, K., Leibe, B., Schiele, B.: Multiple object class detection with a generative model. In: CVPR, vol. I, pp. 26–36 (2006)

22. Ohba, K., Ikeuchi, K.: Detectability, uniqueness, and reliability of eigen windows for stable verification of partially occluded objects. PAMI **19**(9), 1043–1047 (1997)

23. Okada, R.: Discriminative generalized hough transform for object dectection. In: ICCV, pp. 2000–2005 (2009)

24. Russakovsky, O., Ng, A.: A steiner tree approach to efficient object detection, pp. 1070–1077 (2010)

25. Schneiderman, H., Kanade, T.: Object detection using the statistics of parts. IJCV **56**(3), 151–177 (2004)

26. Stauffer, C., Grimson, W.E.L.: Adaptive background mixture models for real-time tracking. In: CVPR, pp. 2246–2252 (1999)

27. Viola, P.A., Jones, M.J.: Robust real-time face detection. IJCV **57**(2), 137–154 (2004)

28. Wang, Z., Kagesawa, M., Ono, S., Banno, A., Ikeuchi, K.: Global 3d modeling and its evaluation for large-scale highway tunnel using laser range sensor. In: ITS World Congress (2012)

29. Wojek, C., Roth, S., Schindler, K., Schiele, B.: Monocular 3d scene modeling and inference: understanding multi-object traffic scenes. In: ECCV, vol. IV, pp. 467–481 (2010)

30. Xue, L., Ono, S., Banno, A., Oishi, T., Ikeuchi, K.: Global 3d modeling and its evaluation for large-scale highway tunnel using laser range sensor. In: ITS World Congress (2012)

31. Yarlagadda, P., Monroy, A., Ommer, B.: Voting by grouping dependent parts. In: ECCV, vol. V, pp. 197–210 (2010)

32. Yeh, T., Lee, J., Darrell, T.: Fast concurrent object localization and recognition. In: CVPR, pp. 280–287 (2009)

Chapter 7

User Needs in Green ITS: Results of a Questionnaire Survey and Proposal for Green ITS Design

Wing Yan Man[1], Jing Bie[2], Bart van Arem[3]

[1]Department of Civil Engineering, Faculty of Engineering Technology, University of Twente, Drienerlolaan 5, 7522 NB Enschede, The Netherlands

[2]Centre for Transport Studies, Faculty of Engineering Technology, University of Twente, Drienerlolaan 5, 7522 NB Enschede, The Netherlands

[3]Department of Transport and Planning, Faculty of Civil Engineering and Geosciences, Delft University of Technology, 2628 CN Delft, The Netherlands

ABSTRACT

An internet survey has been conducted among drivers in the Netherlands and Japan to study their attitude towards Green ITS. The survey focuses on driving behavior and ITS experience, attitude towards environment, and Green ITS preferences. The results show that money-related information gives the most effective motivation for drivers to follow fuel efficient behavior; the least preferred information is impact on environment. Correlations are found between information preference and the driving style. The majority of the respondents intend to make frequent use of Green ITS. Based on these results, a new design for Green ITS is proposed, providing drivers with information on driving behavior, expenses on fuel consumption and environmental factors.

KEYWORDS

Green ITS, User needs, Fuel efficiency, Questionnaire survey, User interface

1. INTRODUCTION

Excessive energy consumption and greenhouse gas (GHG) emissions are making changes to our living environment that will be damaging in the long term. A prominent example is global warming, as acknowledged in the Kyoto Protocol [1]. Road traffic has been a major contributor to such changes. Now, more than ever, people are being encouraged to act upon it to create a better environment for the future. Diminishing car use would decrease GHG productions [2] but this is difficult to achieve in this modern and dynamic world [3]. A realistic solution would be to improve fuel efficiency, such as to guide our drivers to a fuel efficient driving behavior. High speeds, heavy acceleration and lack of anticipation in traffic have been identified as the main causes of excessive emission production [4]. By improving the driving skills to a more environmentally friendly style, we can reduce fuel consumption, fuel expenses as well as emission production. This benefits both the individual and the environment.

Green ITS (Green Intelligent Transport System) monitors the driver's performance in order to achieve a more fuel efficient behavior. By displaying information on driving history and CO_2 emission, the system informs drivers of their impact on the environment. Nowadays, there are a few systems available on the market that monitor driving behavior, such as Fiat eco:Drive [5] which reports on the fuel economy after the trip. In other systems, fuel consumptions are shown to the driver dynamically. To the best of our knowledge, no user feedback has been published in terms of driver response and behavioral adaptation due to Green ITS. To promote and improve these systems, the design and functioning should satisfy the desires of the user [6]. By measuring the attitude of drivers towards Green ITS, designers will be aware of the needs and preferences of drivers. Green ITS can then be designed from both a technological point of view and the preferences of the driver.

This research focuses on user needs in Green ITS. The objective is to improve current designs of Green ITS by considering factors that motivate drivers to change their behavior. Based on the results of an internet survey, this paper will elaborate on user attitude and preferences towards Green ITS among Dutch and Japanese drivers. A new design of Green ITS is then proposed which motivates drivers to follow environmentally friendly driving behavior.

2. Background

To achieve sustainable mobility, various measures are being implemented to reduce GHG emissions around the globe. Traditional measures include increasing capacity of the road network (such as constructing extra lanes) and imposing fuel tax for transport [7]. The car industry has been active in introducing electric/hybrid cars and information technology systems that support the driving task [8]. For reducing GHG emission, researchers have been focusing on engine technologies, policy measures, changes to the infrastructure, modal shift, transport management, and driving behavioral changes.

In this paper we focus on the driving behavioral changes. For most drivers it is not easy to change their current driving behavior by adopting new skills. To help the driver in changing his (or her) behavior, in-car Green ITS systems and services are being developed. These systems aim at achieving a fuel efficient behavior. They provide the driver with information on his driving behavior: the driver is made known of his performance by several indicators, displayed during and/or after the trip. Examples of the indicators include characteristics of fuel consumption and speed profile. More advanced systems collect driver behavior data from previous trips and compare them with the optimal fuel efficient behavior. Personal advices can then be provided to the driver to achieve the optimal fuel efficiency. It is also possible that these advices are given in real time (i.e. during the trip, while the driver is driving). Such a system monitors the current driving behavior and intervenes if inefficient behavior is being practiced. The driver receives direct feedbacks on his behavior (progression) and advices on changing his behavior for better fuel efficiency.

3. QUESTIONNAIRE SURVEY

For a successful design of in-car Green ITS systems and services, user needs and acceptance are essential. We have deployed a questionnaire survey to investigate drivers' opinions towards Green ITS. The purpose of this survey is to assess the following aspects:

- Type of preferred information for motivating the driver to change into fuel efficient behavior;
- Influence of driver's personal characteristics on the type of preferred information;
- Attitude towards Green ITS.

The currently available in-car technologies are mostly designed from a technological point of view. With the knowledge of driver attitude and preferences, we can then improve the current designs of Green ITS.

3.1 Questionnaire Design

The questionnaire survey "Green light for a better environment" has been deployed and it consists of questions on the following four subjects:

- General information;
- Driving behavior & ITS experience;
- Attitude towards environment;
- Attitude towards Green ITS.

The survey first provides the respondent with a brief introduction on Green ITS. This introduction does not include the detailed attributes of Green ITS (such as the possibility of different interfaces, sounds and lights), in order to avoid confusion and pre-selection bias.

3.1.1 General Information

This part consists of questions on the personal characteristics of the respondent. Only respondents possessing a driving license are asked to participate in the survey. The questions are related to age, occupation, type of driving licenses, vehicle ownership, and vehicle usage (frequency and vehicle features). With this information drivers can be categorized. Factors such as driving experience and personality can also be easily retrieved from the dataset.

3.1.2 Driving Behavior & ITS Experience

Questions in this part are related to average mileage, driving frequency, trip purposes, route choice, and familiarity with ITS. By investigating the respondent's driving behavior we can identify which driver group yields the highest impact on the environment. This driver group needs special support from Green ITS. Also, driver experience with ITS is studied to assess driver familiarity with ITS technologies, which influences driver adaptability towards Green ITS. This is an important factor in the design of Green ITS.

3.1.3 Attitude Towards Environment

Drivers are asked to rank several types of information in terms of the likelihood to trigger a behavioral change on them. Other questions relate to the attitude towards environment in general and the self image of the respondent on environmental friendliness. By measuring driver attitude towards the environment we can identify the type of drivers who have a positive attitude and are therefore more willing to change their behavior for a better environment. Drivers who are not impressed by the environmental issues need other stimulus for changing their behavior.

3.1.4 Attitude Towards Green ITS

To design a Green ITS system that satisfies the desires of the driver, we need to measure the driver's attitude. Many drivers are not yet familiar with in-car Green ITS, because it is still a novelty. To measure their attitude towards such systems, examples of in-car Green ITS are provided to give the respondents a reasonable impression. Respondents are then asked which type of information is relevant to help them in adapting their behavior. Four categories of information are considered:

- general information such as fuel consumption, costs and CO_2emission;
- advice on driving behavior;
- analysis of previous driving behavior;
- progress in driving behavior.

Besides the type of information, the timing of information provision is also studied.

3.2 Conducting the Survey

Internet survey has been chosen as survey method. It has the advantages of easy accessibility, low costs, direct collection of digital data, and convenience in data processing. Internet surveys provide an effective way of reaching the drivers and collecting their opinions. In this study we only consider drivers from the Netherlands and Japan. As of June 2010, 88.3% of the whole population in the Netherlands are internet users. The same statistic is 78.2% for Japan [9]. We expect the ratio of internet users among Dutch and Japanese drivers is around or even higher than these percentages. Therefore internet surveys do not suffer greatly from the selection bias of excluding non-internet users.

The survey is set up in both the Dutch and Japanese languages. Limeservice (www.limeservice.com) has been chosen as the survey interface owing to its ability to edit both Latin and Japanese characters. The final web survey is hosted on the following web pages:

- ***Dutch version:*** http://wingyan.limeask.com/index.php?sid=56984&lang=nl
- ***Japanese version:*** http://wingyan.limeask.com/index.php?sid=97472&lang=ja

The survey was conducted in the period between December 2008 and February 2009. The respondents were informed of the survey by an email which included information on the research and a link to the survey webpage. Respondents in the Netherlands were reached through a personal network. They were also asked to forward the questionnaire among their social networks. To reach the Japanese respondents, the survey was spread over companies and universities. This snowball method turns out to be quite effective, due to the efficiency in email services and the flexibility of a web survey.

4. DATA ANALYSIS

In total 100 respondents from the Netherlands and 98 respondents from Japan filled in the survey. Table 1 shows the general characteristics of the respondents. For both countries the majority of the respondents are between 18 and 39 of age, and students in profession. There is no professional driver respondent from Japan. The variation in profession is higher in the Netherlands than in Japan.

Between the respondents from these two countries, significant differences are observed in terms of car possession, annual average mileage and primary trip purpose. Some Dutch drivers possess more than one car, while most of the Japanese drivers surveyed have one car or no car. Dutch respondents have a higher average mileage, make more frequent use of the car, and are more likely to commute by car. On the contrary, Japanese respondents have a lower mileage, use the car occasionally and mainly for private situations.

These differences are attributed to the different demographics in these two countries. Urban population density is higher in Japan than in the Netherlands. Ridership of public transport is also much higher in Japan. Using car to commute is far less popular in Japan than in the Netherlands. Japanese drivers often use the car for more private purposes, such as social and recreational activities.

The differences are also partially attributed to the unbalanced representation of the population within the respondents. As noted above, young people and students make out a big proportion of the respondents. The group of the respondents surveyed in this study is therefore an unrepresentative sample of the society. This bias will be taken into account in our analysis below.

Table 1. Characteristics of respondents

Characteristics		Dutch	Japanese
Age	18–24	48%	40.6%
	25–39	36%	38.5%
	40–64	13%	19.8%
	65 and older	3%	1.0%
Profession	Student	39%	53.1%
	Office worker	28%	30.2%
	Professional driver	5%	0.0%
	Other	28%	16.7%
Car possession	None	32%	44.8%
	1	38%	45.8%
	2–3	28%	9.4%
	More than 3	2%	0.0%
Annual mileage	0–1.000 km	11%	41.7%
	1.000–5.000 km	19%	29.2%
	5.000–10.000 km	13%	14.6%
	10.000–20.000 km	29%	12.5%
	20.000–30.000 km	13%	1.0%
	More than 30.000 km	15%	1.0%
Primary trip purpose	Commuting	41%	15.6%
	Work related	13%	3.1%
	Social & Recreational	45%	56.3%
	Shopping	1%	25.0%

4.1 Information Priority

Instead of categorizing drivers by their personal characteristics and then examining per group the preferred information, we have done the analysis in reverse. In the survey the respondents are asked to select the most relevant information for them to receive in order to change their behavior; they had to rank four types of information by priority. The results provide the reference point in our analysis: this priority information defines the respondent's character and attitude towards environment. By correlating this character to his driving style and experience with ITS, suitable Green ITS based on the preferred information can be designed.

Table 2 shows the frequencies of the four types of information being ranked with the highest priority. For both countries, money related information is the most important to drivers. The second most preferable information is about fuel consumption; this information is more preferable for Japanese drivers than for Dutch drivers. A remarkable difference is the importance of CO_2 emission between the two countries.

Table 2. Distributions of priority information

Priority information	Dutch	Japanese
Fuel consumption	17%	35%
Money saving	64%	56%
CO_2 emission	11%	2%
Contribution to environment	8%	7%

4.1.1 Personal Character

Several key factors have been chosen to uncover correlations between the preferred information and the character of the driver. To define the personal character of the driver, three key factors have been analyzed: age, profession and car possession. Age and profession tell something about the driver's daily life. Generally, office workers drive or commute more by car than students, which makes them a more important target for Green ITS. Car possession indicates the driver's need of a car and/or desire to drive. These factors are then compared with the preference of information.

The survey results show that Dutch respondents between the age 18 and 24, being students, and without car possession, are much more interested in money related information than other driver groups. The fact that office workers are forced to use the car and usually can declare their expenses to their employer explains the importance of fuel consumption instead of expenses. For the Japanese respondents, no correlation between choice of priority information and personal character has been identified. Each driver group (by age, profession or car possession) exhibits a distribution of priority information similar to the overall distribution in Table 2.

4.1.2 Driving Style

Driving style characterizes the driver's driving behavior in daily life. The annual mileage and driving frequency provides information on car use. Drivers with a higher mileage produce more emissions. Primary trip purpose reveals the driver's familiarity with the routes during an ordinary trip. Commuting is often via the same route, meaning that the driver is familiar with the route and more likely to drive in a smooth way. This leads to less emission compared to recreational trips, where the driver is unfamiliar with the routes and has to learn about the conditions on these routes.

The survey results show that Dutch respondents with an average to low annual mileage (1,000 km to 10,000 km) and less frequent use of car (1 to 8 times per month) are much more interested in money savings than other driver groups. No major difference has been identified between trip purposes. For the Japanese respondents, it appears that annual mileage and driving frequency have very little influence on the choice of priority information. However, respondents with their primary trip purpose as private (social recreational and shopping) are slightly less interested in money savings than other driver groups.

4.1.3 Experience with ITS

Drivers' previous experience and familiarity with ITS systems and services make it easier for them to adapt to Green ITS, if the Green ITS is designed in a comprehensible way and in accordance with other ITS systems. In this survey, respondents are to rate their familiarity with cruise control, adaptive cruise control, lane departure warning, navigation systems, and intelligent information systems.

The survey results show a striking difference between the two countries in terms of the respondent's familiarity with ITS systems. Japanese respondents appear to be significantly less familiar with current ITS systems than the Dutch respondents, possibly due to the large amount of students in the respondent group.

Nonetheless, cruise control and navigation systems are widely used and perceived as familiar in both countries. Therefore a comprehensible new system should have the same level of adaptability as these two systems. For example, some interface aspects of these systems can be borrowed to maintain familiarity with the new system.

However, familiarity with ITS is shown to be not strongly related to the choice of priority information. Among drivers who are experienced with a particular ITS system, similar ratios in information preference are observed as those in Table 2.

4.1.4 Attitude Towards Environment

Environmentally friendly behavior concerns actions that people take to improve the environment, such as recycling or low energy use. Most respondents hesitate to call themselves very environmentally friendly. The majority of respondents reflect on themselves as having an average personality towards environmental friendliness. Dutch respondents with CO_2 emission and contribution to environment as their priority information do think that they are more environmentally friendly. On the contrary, Japanese respondents who are interested in information on their contribution to environment mostly refer to themselves as not environmentally friendly.

The respondents are also asked to reflect on their own fuel consumption behavior during driving. Respondents who choose contribution to environment as their priority information generally consider their driving behavior as not very fuel efficient, indicating their need to acquire such information. On the other hand, respondents who consider themselves as fuel efficient mainly choose fuel consumption as their priority information, signifying their desire to further reduce their fuel consumption.

Another question in the survey is about the influence of fuel price on car use. The intention of raising fuel price is to reduce fuel consumption. It is expected that respondents with fuel consumption and money savings as their priority information would be highly affected by fuel price. However, this is not the case in the survey. For the Japanese respondents, the effect of raising fuel price is independent of the driver's priority information; for each of the four respondent groups (as defined by Table 2), about half of the respondents report fuel price as having an influence while the other half state no influence. For the Dutch respondents, raising fuel price has limited influence on their behavior. For those with money savings as their priority information, the majority of them actually claim that fuel price has no influence at all on their behavior. A possible explanation is that these drivers are dependent on their car for commuting and cannot reduce their frequency of car use even under high fuel prices.

4.2 Relevant Information for Changing Driving Behavior

In the survey respondents are asked to rate several types of Green ITS information in terms of relevance. A high relevancy is interpreted as a high willingness to use Green ITS and to adapt one's driving behavior according to the received information. Information with the highest relevancy is the most preferable for the drivers to receive. Next to the types of information, respondents are also asked to indicate their preferred timing of receiving the information (discussed in section 4.3).

4.2.1 Monetary Impact

The majority of drivers chose money related information as the most relevant motive for them to change their behavior. These drivers also include the most frequent drivers, i.e. those who produce the most emissions and with whom the most impact of Green ITS can be achieved. The information can be directly or indirectly related to money, including fuel consumption, money converted from fuel consumption, and the amount of money saved with fuel efficient behavior (Fig. 1). This type of information is much more relevant for Japanese drivers than for Dutch drivers, although more Dutch respondents have chosen money related information as their priority information than Japanese respondents.

4.2.2 Environmental Impact

As for the types of environment related information, the indicator for fuel efficiency is the most popular among both Dutch and Japanese drivers (Fig. 2). For Dutch drivers, information on CO_2 emission and contribution to environment are in general irrelevant, although the number of drivers who do indeed rate such information as relevant is not negligible. However, Japanese drivers generally find environment related information as relevant and are willing to receive such information. A possible reason is curiosity, i.e. the eagerness among Japanese drivers to use new technology and to receive all kinds of information.

4.2.3 Personal Driving Behavior

Five elements of personal driving behavior are considered: speed, gear shift, acceleration, deceleration, and route choice. Information related to such behavior includes the advice on optimal behavior, the en route reporting of driving progress, and the post trip reporting of driving history. Both Dutch and Japanese drivers find advices on their behavior very relevant (Fig. 3). Moreover, Japanese drivers show an equal interest in receiving information on their driving progress and history, whereas Dutch drivers find such information less interesting than advices. In terms of the behavioral elements, Dutch drivers prefer to receive information on speed and gear shift than on acceleration and deceleration, while Japanese drivers are indifferent.

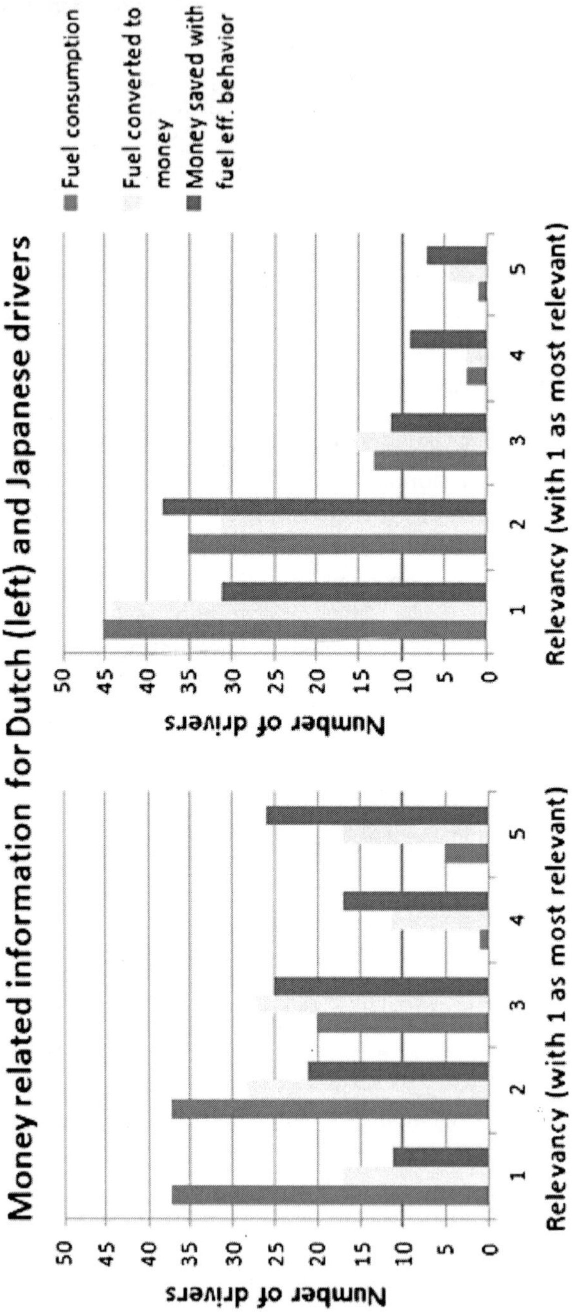

Figure 1. Level of relevance for money related information: Dutch drivers (*left*) and Japanese drivers (*right*)

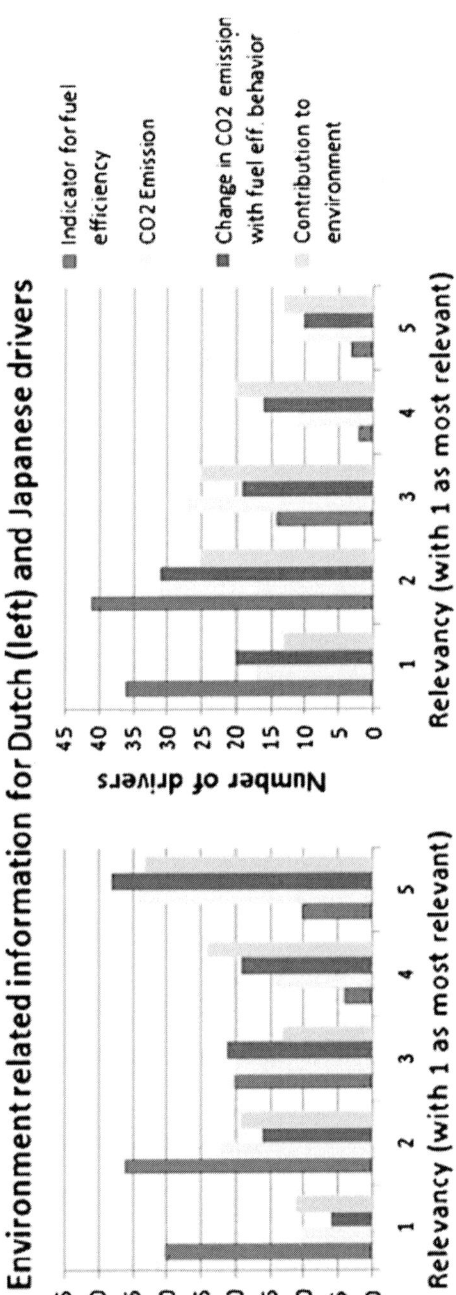

Figure 2. Level of relevance for environment related information: Dutch drivers (*left*) and Japanese drivers (*right*)

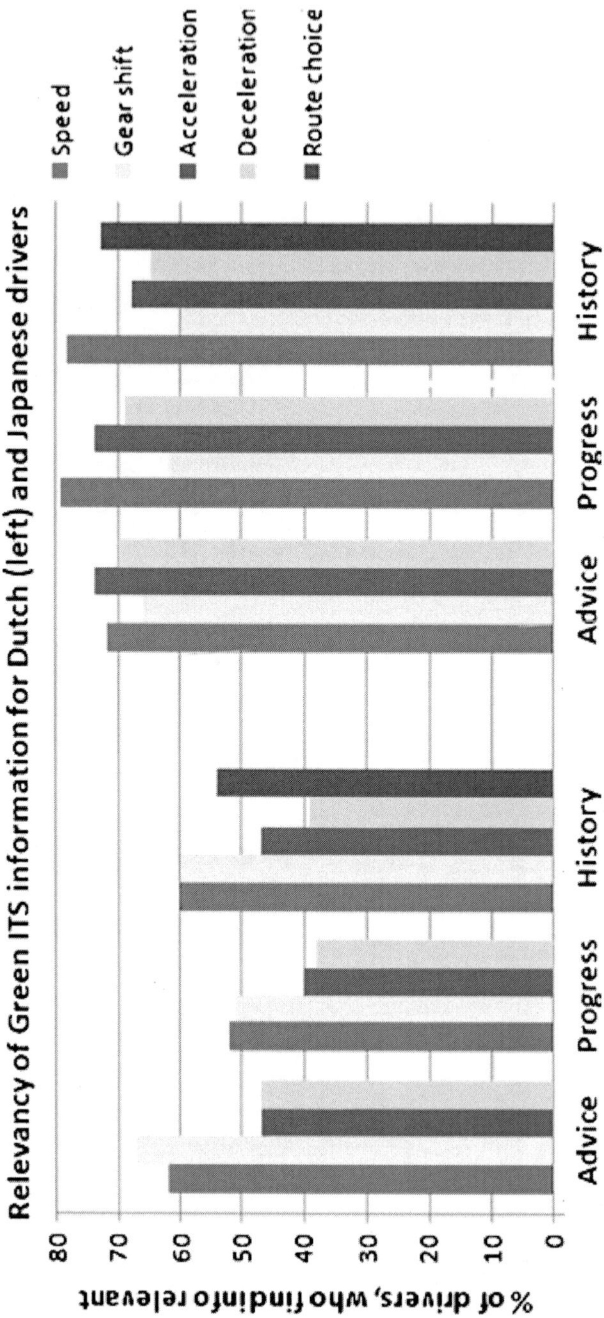

Figure 3. Level of relevance for different types of Green ITS information: Dutch drivers (*left*) and Japanese drivers (*right*)

4.3 Timing of Information

In the design of Green ITS, timing of the information supply is very important. Wrong timing of information can reduce the impact on driving behavior. An example is when advice on behavior is only given after trips. After trips drivers usually do not care about their behavior anymore and for the next trip they are more likely to have forgotten the advice.

In the survey respondents are presented with control-related advices and money-related information. They are then asked for their preferences on the timing of these advices and information. In terms of control-related advices, both Dutch and Japanese drivers prefer to receive such information during the trip; this preference is especially strong among Dutch drivers. In terms of money-related information (Fig. 4), Dutch drivers prefer to receive information on fuel consumption during the trip and information on money savings after the trip. Japanese drivers, however, prefer to receive both types of information during the trip.

5. PROPOSAL FOR GREEN ITS DESIGN

Based on results of the internet survey, improvements can be made to current Green ITS systems to better accommodate drivers' preferences. Accordingly, a new Green ITS design is proposed here.

5.1 Data Flow

History, progress and advice on driver behavior are based on the personal driving behavior, i.e. the factors of speed, gear shift, and ac-/deceleration. During the trip, data of these factors are collected and then compared to the optimal fuel efficient behavior (e.g. optimal speed, optimal gear shift moment) in order to determine the advice to be given to the driver. Once the driver is following the advice, the difference due to this behavioral change is calculated.

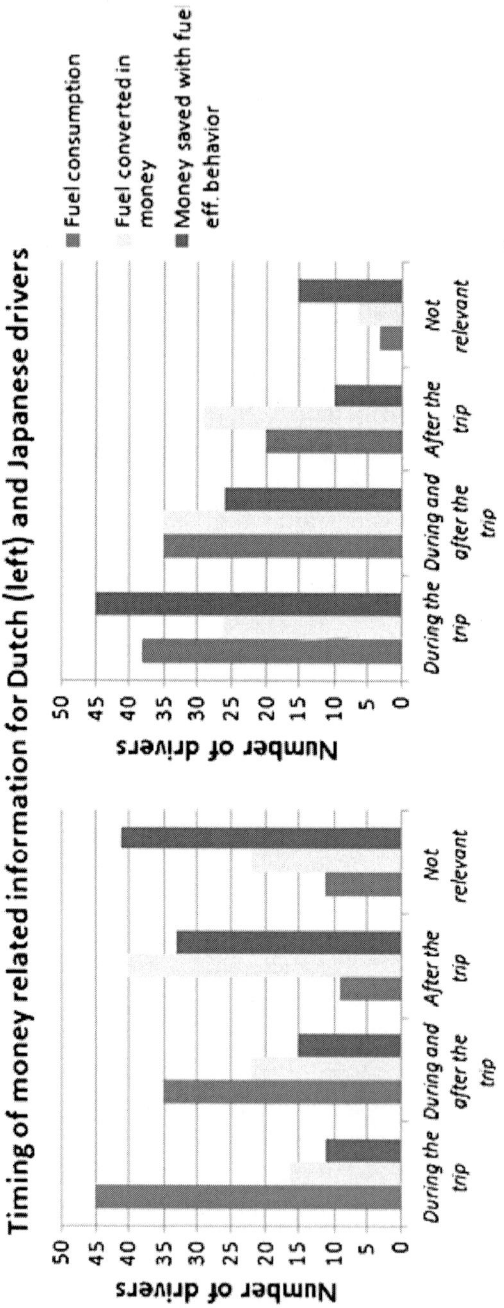

Figure 4. Preferred timing of the money related information: Dutch drivers (*left*) and Japanese drivers (*right*)

Besides information on personal driving behavior, other types of information are also available to the driver. Money related information includes fuel consumption in liters per kilometer, fuel consumption calculated in monetary values, and the amount of money savings in fuel consumption (compared to the "previous" behavior). Environment related information includes the amount of CO_2 emission and the reduction of CO_2 emission.

5.2 Type of Information

The use of Green ITS should not affect the driver's safety. To avoid stress, distraction and confusion during driving, drivers can choose to only display the essential information directly related to their driving behavior such as progress of the personal driving behavior. Other types of information can be displayed before and/or after the trip. Information that has direct effects on the driving behavior is advice for fuel efficient behavior. Unlike information about driving progress and history, advice can be directly adapted by the driver. The driver has the choice to display advice for speed, gear shifts, and ac-/deceleration.

For money related information, it is suggested to only display fuel consumption during the trip. Because of changing fuel prices, monetary values are not always accurate and are better displayed before or after the trip (with the driver being able to specify the fuel price). All environment-related information can be made available after the trip.

5.3 User Interface

Figure 5 gives an impression on the proposed Green ITS system. During the trip the information displayed should be easily understandable and without any confusion or distraction. An on board unit is attached to the dashboard with its screen showing information on gear shift, and ac-/deceleration. The screen will gradually turn red when inefficient behavior is detected, such as excessive pedal pressure and gear shift at high revolutions. The deeper the color, the more inefficient the behavior is. With fuel efficient behavior the display turns into green again. The bar in the corner displays how much fuel is left. The factors of speed, gear shift, and ac-/deceleration are displayed in line charts on screen, showing both the actual behavior (the red curve) and the optimal behavior (the green curve).

Figure 5. Conceptual sketch of the proposed Green ITS system

Graphs displayed after the trip provide overviews on the differences between the actual fuel consumption and the fuel consumption under fuel efficient behavior (in both volume and money). This tells the driver how much excessive fuel has been consumed and how much money could have been saved with fuel efficient behavior. The same concept applies to environment related information, i.e. differences between the actual CO_2 emission and the emission under optimal behavior.

6. CONCLUSIONS AND DISCUSSIONS

Through an internet survey we studied the attitude towards Green ITS among drivers in the Netherlands and Japan. It has been shown that the biggest motivation for drivers to follow fuel efficient behavior is the potential money savings. Green ITS should strive to provide drivers with information on how much money they would save. Next to that, drivers are also interested in information on fuel consumption during their trips. The least preferred information is actually about the impact on environment, such as CO_2 emission. The cause of this disinterest could be that drivers do not see a direct relationship between their car use and its impact on the environment. In order to make these drivers aware of their impact on environment and to motivate them to adapt to environmentally friendly behavior, we need to help the driver establish a direct, understandable and imaginable relationship between their behavior and the environmental impact. An example could be to make a comparison between emission and the reduction of trees.

Personal characteristics, experience with ITS, and attitude towards environment have been shown to cast only a minor influence on the driver's preference of information in Green ITS. Correlations, however, have been found between the preference of information and the driving style. Overall, the majority of respondents consider information provided in Green ITS as relevant and they are willing to use such systems. This positive attitude is expected to contribute to high acceptance and frequent use of Green ITS. However, the limited scope of this study should be noted in future user preference studies where more representative respondent sampling should be made. It should also be noted that drivers' actual preference and behavior (revealed preference) may turn out to be different than what they have stated in the questionnaire survey (stated preference).

Green ITS is a relatively new concept. Very limited deployment experiences have been reported in literature. For future implementation and promotion of Green ITS, information on driver experiences should be gathered. By analyzing these experiences, likes and dislikes of users can be identified, which will make room for further improvement. These improvements can be based on the type of information supply, the timing of information supply and the display of information. Different types of information or modes of display are expected to have different effects on the driving behavior.

In the design of Green ITS, safety, distraction and other psychological influences on the driver are of great importance. In the proposed design, the system provides information on driving behavior, expenses on fuel consumption and environmental factors. During the trip, advice and fuel consumption will be displayed with simple indications on an on-board screen. After the trip other information will be presented with trends, numbers and texts.

ACKNOWLEDGMENTS

The research described in this paper was conducted during the first author's bachelor assignment at Toyota InfoTechnology Center Co., Ltd. The authors would like to thank Sachi Mizobuchi and the Department of Research and Development for their enthusiasm and contribution to this research. The authors also acknowledge the constructive comments of an anonymous referee.

REFERENCES

1. UNFCCC, Kyoto Protocol. Available online: http://unfccc.int/kyoto_protocol/items/2830.php (1997)
2. Logghe, S., Van Herbruggen, B., Van Zeebroeck, B.: Emissions of Road Traffic in Belgium. Transport & Mobility Leuven (2006)
3. Fergusson, M., Smokers, R., Passier, G., ten Brink, P., Watkins, E., Valsecchi, C., Hensema, A.: Possible regulatory approaches to reducing CO2 emissions from cars. European Commission, Brussels (2007)
4. Bell, M.C.: Environmental factors in intelligent transport systems. IEE Proceedings: Intelligent Transport Systems **153**, 113–128 (2006)
5. FIAT, eco:Drive. Available online: http://www.fiat.co.uk/ecodrive/ (2009)
6. van Driel, C.J.G.: Driver Support in Congestions. TRAIL Research School, Delft (2007)
7. van Breugel, P.B., Baum, A., Calovi, L., Hackman, M., de Gier, C.W., Juneholm, M., Klaeboe, R., Pucher, E., Kampfer, A., Vinot, J.: Examples of air quality measures near roads within Europe: National measures of the international CEDR air quality group. Rijkswaterstaat, Delft (2005)
8. van Maarseveen, M.F.A.M., van der Voort, M.C., van Driel, C.J.G.: Qualitative Survey on Fuel Economy Devices. University of Twente, Enschede (2002)
9. de Argaez, E., Internet World Stats. Available online: http://www.internetworldstats.com/ (2011)

Chapter 8

Incident Detection in Heavy Traffics in Tunnels by the Interlayer Feedback Algorithm

Shunsuke Kamijo , Kaichi Fujimura

Institute of Industrial Science, The University of Tokyo,
4-6-1 Komaba, Meguro-ku, Tokyo 153-8505, Japan

ABSTRACT

In this paper, we developed video surveillance system to detect incidents from heavy traffics in tunnels. Incidents in tunnels may cause additional accidents or fires that may result in fatal disaster. Therefore, it is important to detect incidents as soon as possible to manage the traffics inside the tunnels by the officers. Generally, video images in the tunnels suffer from heavy occlusion due to the low position of camera settings. In particular, the problem of heavy occlusions would be more serious in the urban tunnels due to their heavy traffics. We developed a tracking algorithm to segment vehicles and estimate the precise vehicle trajectories against the heavy occlusions. Utilizing this tracking algorithm, dedicated algorithm to detect incidents from the traffic images was developed. For the experiments, video streams of three cameras for 6 months were investigated, and 32 incidents were examined to evaluate the developed algorithm.

KEYWORDS

Incident detection, Tracking, Occlusion, Segmentation, Semantic hierarchy

1. INTRODUCTION

There are a lot of previous works on the vehicle tracking. Historical works of the vehicle tracking are as follows. A work by H. Kolling and H. Nagel [1] is widely known as successful by using vehicle shape model consisting of horizontal and vertical edges. Peterfreund [2] developed boundary extraction method by 'Snakes [3]' for vehicle tracking. Smith [4] and Grimson [5] employed statistical analysis and clustering of optical flows for the tracking problem. Although those works do not deal with occlusion problem, some previous works deal with the problems. Leuck [6] and Gardner [7] employed 3D models of vehicle shape and developed a dedicated pattern matching algorithm to vehicle images. However, since this method requires various kinds of vehicle models, it would be difficult to be applied to a scene containing various kinds of vehicle traveling into various directions. Malik [8] tried to resolve occlusion problem by a dedicated reasoning method. Although, those methods achieved good performance in a sparse traffic with simple motion, they cannot deal with complicated traffics.

In recent years, the tracking algorithm has been restlessly improved [9–13]. Among them, Kanhere [10] tried to solve the occlusion problem at low camera angle by employing a 3-D perspective mapping from the scene to the image along with a plumb line projection. Zhang [12] applied multilayer framework to the occlusion problem. In this work, occlusion is detected by evaluating the compactness ratio and interior distance ratio of vehicles, and the detected occlusion is handled by removing a "cutting region" of the occluded vehicles. Ambai [13] developed the tracking method against occlusions consisting of feature point detection and clustering feature points. Gradient based feature detector was used for the feature point detection, and graph cut method was used for dividing the occluding objects.

A lot of researches on incident detection have been performed. Cuchira [14] applied traffic rule-based reasoning to images for simple traffic conditions on a single-lane straight road. Jung [15] tracked vehicle from traffic images for precise traffic monitoring. Zhang [16], Srinivasan [17], and

Byun [18] employed Bayesian Network and Neural Network in order to distinguish incident from ordinary traffics. All the above algorithms do not consider about the effect of occlusion. Thus, they have some difficulties in being applied to complicated traffic situation such as congestion. Thomas [19] and Zheng [20] tried find outliers in the traffic flow statistics for the incident detection. Although their methods were successful, the methods require longer time for the detection than video analysis. Shehata [21], Versavel [22], and Austin [23] employed video surveillance for the incident detection. Although their methods were successful, they are focusing on relatively easier cases such as the detection of vehicles stalling at prohibited areas.

Incident detection in the heavy traffic is quite challenging because of heavy occlusions. The incident detections in the tunnel become more difficult due to the low angle condition of the camera settings. In this paper, we resolved the heavy occlusion problems by the interlayer feedback strategy, and developed a successful algorithm to detect incidents even in the congested traffic in the tunnel. The algorithm was examined using video images recorded in the tunnel at the Tokyo Metropolitan Expressway.

2. ALGORITHM HIERARCHY FOR SCENE UNDERSTANDING

Semantic Hierarchy is one of popular idea to construct cognitive systems. In this section, an algorithm hierarchy for traffic scene understanding is discussed in detail. In particular, an innovative idea of interlayer feedback algorithm was developed in this paper, and the idea will be described in the last part of this section. Although the discussions exemplify traffic scene understanding, the idea would be useful for cognitive systems in general.

2.1 Overview of Semantic Hierarchy

Generally, algorithm hierarchy for cognitive systems can be separated into two groups of layers, Signal Processing Layer and Semantics Layer (Fig. 1). Signal Processing Layer is responsible for transforming the electronic signal such as image data into designated data having significance in the real-world such as vehicle trajectories. Semantics Layer is responsible for transforming data obtained from Signal Processing Layer into semantic symbols verbally recognizable by human.

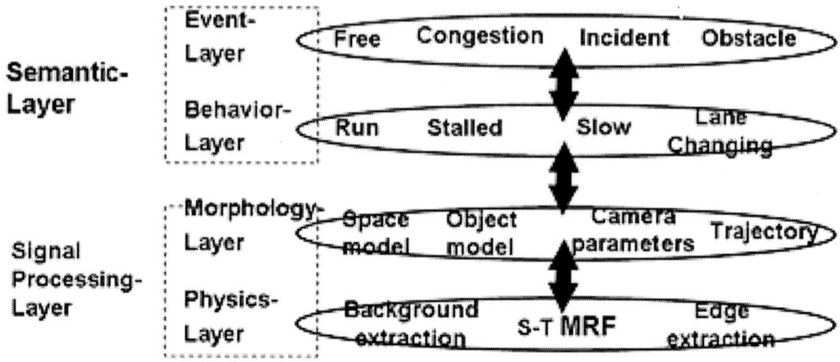

Figure 1. Algorithm hierarchy for the traffic surveillance

Physics Layer in Signal Processing Layer should contain algorithms that refer purely intrinsic information within images themselves. In other words, the algorithms in Physics Layer should not refer to any information such as forms of vehicles and human, coordinate geometry between the image and the real world, and topology of vehicle lanes. Typically, the methods for background image acquisition or edge extraction are considered as Physics Layer algorithms. Thus, such the algorithms can be generally applied to a variety of objects and scenes. On the other hand, the algorithms in Morphology Layer should refer to external information that cannot be obtained from images themselves. Models for vehicle shapes or human bodies, real world geometry, camera parameters, and vehicle lane topology are examples for such the external information.

Previous works on object tracking does not usually distinguish Physics Layer and Morphology Layer. However, it is important to clarify the difference of the two layers and to combine them appropriately, in order to maintain both benefits of the generality of Physics Layer and the respectiveness of Morphology Layer. In this paper, the idea of interlayer feedback will be discussed in detail as can be read in Section 3.

In Behavior Layer of Semantics Layer, vehicle trajectories obtained from Signal Processing Layer are transformed into verbally significant symbols that represent each vehicle's behavior such as stalling, slow, and changing lanes. Finally, algorithms in Event Layer determine the most preferable event to represent the traffic scene by combining behaviors of vehicles obtained from Behavior Layer.

As an example, when the system detected a stalling vehicle and the following vehicles changing lanes to avoid the stalled vehicle frequently, it should be determined as a scene of an incident. On the other hand, even when the system detected a stalled vehicle, it should be determined as congestion if the following vehicles do not move to the same way.

2.2 The S-T MRF Model as a Physics Layer Algorithm

The Spatio-Temporal Markov Random Field model [24, 25] was developed in order to segment spatio-temporal images by referring to purely intrinsic information within image sequences. It divides an image into blocks as a group of pixels, and optimizes labeling of such blocks by referring to texture and labeling correlations among them, in combination with their motion vectors. Combined with a stochastic relaxation method, the S-T MRF optimizes object boundaries precisely against occlusions.

For example, Object-map representing segmentation result map of object ID distribution will be obtained as shown in Fig. 2. This result is obtained by applying the S-T MRF model to the image from very low-angle at merging traffic on an expressway. In the tracking images of Fig. 2a, bounding boxes overlaps each other. However, vehicles were segmented correctly and the overlapped rectangles are just a problem of visualization as appeared in the Object-map in Fig. 2b.

Since the S-T MRF model was defined as an algorithm in Physics Layer, it is not able to segment occluded objects moving in similar motions along the traveling. Interlayer feedback strategy would be effective to solve such the occlusion problems with similar motions as described in the following section.

3. INTERLAYER FEEDBACK ALGORITHM

Occlusion problems frequently occur in heavy traffic in the tunnel, and the degrees of occlusions would be serious due to the low angle condition of the camera installation. In addition, occluded vehicles travel to the same direction in the similar velocities in tunnel traffic as shown in Fig. 3a. The S-T MRF model is not able to segment the occluded vehicles with similar motions.

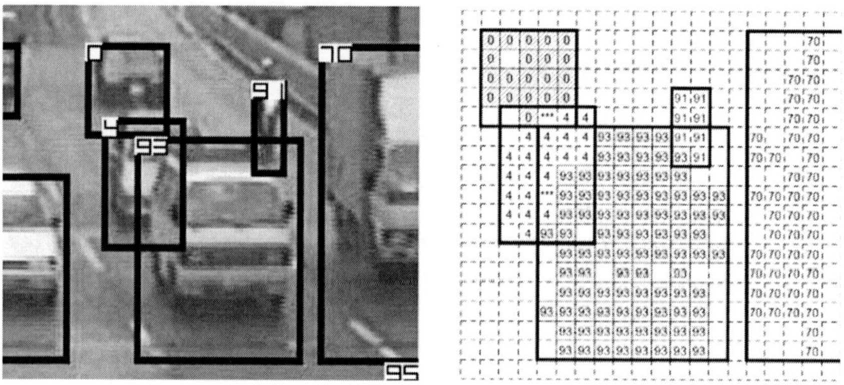

Figure 2. Results from the S-T MRF model

In such cases, algorithms in Morphology Layer should segment the vehicles which were not segmented by the S-T MRF model. For that purpose, we developed a dedicated algorithm to segment occluding vehicles by referring to horizontal edge patterns of vehicles. Vehicle images are rich in horizontal edges, and the patterns are different among the variations of vehicles as sedan, wagon, trucks and buses as shown in Fig. 3b. Therefore, occluding vehicles can be segmented by finding the gaps of horizontal edge patterns. An edge image should be obtained from an original image to find the gap. This is binary edge image defined as a binarized image of illumination invariant edge image used in S-T MRF model [25]. The coordination among S-T MRF model, edge extraction, and this algorithm (Interface-split) is represented in Fig. 4. The Interface-split algorithm is applied to the rectangle judged that two vehicles exist astride two lanes. A description of the algorithm is as follows.

Figure 3. Overview of the interlayer feedback algorithm

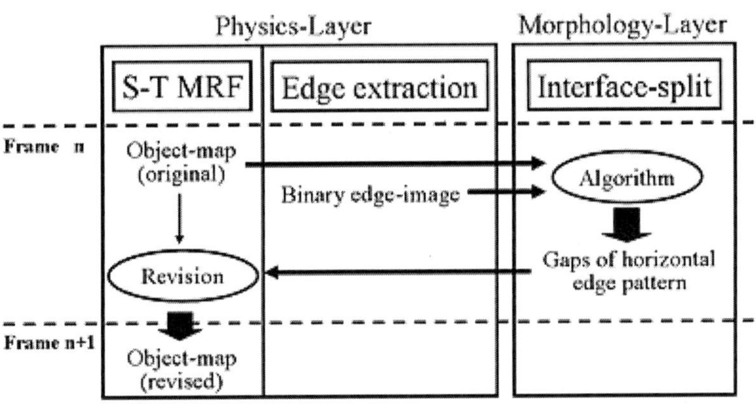

Figure 4. Diagram of interface-split

[Interface-split: detecting gap of horizontal edge]

[Step 1] The scanning line is prepared in a vertical direction of a rectangular image, and the presence of the edge in each block on the scanning line is inspected.

[Step 2] The one-dimensional pattern is generated on the scanning line. This pattern has a binary value corresponding to the existence of edge.

[Step 3] The scanning line is slid to the horizontal direction of the rectangle. A line which has the maximum amount of change in the patterns between the own line and the neighbor line is detected as a splitting line. Hamming distance is used as an index of the amount of change for the line.

[Step 4-a] If the splitting line exists, and if the amount of change for the splitting line exceeds a threshold, object ID on the Object-map is changed based on the splitting line. On the other hand, it is judged that only one vehicle (such as heavy vehicle) exists in the rectangle extracted by the ST-MRF model, and the Object-map is not changed.

[Step 5] In the next frame, the ST-MRF model revised the Object-map based on the result of step 4.

[End of the algorithm]

In this paper, our original method for illumination invariant preprocessing was applied to the original image to cancel the effect of head lights as Fig. 3b, c and dand tracking results in the following sections. The existence of an edge is examined by each block defined in the S-T MRF model, in order to reduce the time and maintain the robustness of the gap finding process.

It is remarkable that after the S-T MRF model is informed segmentation results from the algorithm in Morphology Layer, the S-T MRF model is able to segment the vehicles in the following image frames. Generally, the algorithm in Morphology Layer is not able to perform segmentation and tracking so robustly by itself. Since the S-T MRF model accepts any algorithms in Morphology Layer, the strategy of interlayer feedback is promising to apply to various kinds of objects.

4. OCCLUSION ROBUST ESTIMATION OF VEHICLE TRAJECTORIES

For the incident detection in the tunnel, evaluations of the slow down behaviors and lane changing behaviors of vehicles are necessary. Spatio-temporal estimations of trajectories enable evaluation of vehicle behaviors and scene understanding as incident detection.

In images of low angle cameras in tunnel, a majority of vehicles are not occluded when they come into the view beneath the camera. Therefore, grounding position of the vehicle tail can be estimated from the vehicle segmentation results at the moment when they travel beneath the camera. However, as the vehicles travel toward the downstream in the camera view, tail of the preceding vehicle should be occluded by the following vehicle to cause the failure in estimating grounding position of the preceding vehicle. In this section, a dedicated method to estimate the grounding position of the vehicle tail was developed as described below.

Cameras were originally installed at about 2.5 m high in the tunnel and their optical axes are horizontal. We used them as the original settings. In Fig. 5, y represents vertical coordinate in the image coordinate measured from the vanishing point, and it should be transformed into real world coordinate for the estimation of vehicle velocity in the real world. When the tail of the preceding vehicle is occluded by the following vehicle, y_{grnd} of the preceding vehicle cannot be observed directly. Therefore, y_{grnd} can be estimated from y_{top} and $h_{app}(y_{grnd})$ where y_{top} represents position of the vehicle roof in the image coordinate, and $h_{app}(y_{grnd})$ represents appearance height of the vehicle in the image coordinate when the vehicle exists at the grounding position y_{grnd}. In Eq. 1, since y_{grnd} is estimated from y_{grnd} itself, a feedback optimization process should be employed for the estimation.

$$y_{grnd} = y_{top} + h_{app}\left(y_{grnd}\right) \tag{1}$$

The problem now is to estimate $h_{app}(y_{grnd})$. In Eq. 2, $h_{true}(y_{grnd})$ represents the true height of the vehicle in the image coordinate, and $h_{app}(y_{grnd})$ represents appearance height of the vehicle which consists of roof length and true height of the vehicle as shown in Fig. 6. Figure 6 thumbnails the side view of the road infrastructure, camera setting, and the vehicle. From Fig. 6, the correction from $h_{app}(y_{grnd})$

to $h_{true}(y_{grnd})$ should be done by using correction factor $\Omega\ (y_{grnd})$ as represented in Eq. 2.

$$h_{app}\left(y_{grnd}\right) = h_{true}\left(y_{grnd}\right) \times \Omega\left(y_{grnd}\right) \tag{2}$$

For the purpose of the definition, $\Omega\ (w)$ is function of w, position in the real world coordinate, in Eq. 3. Since $\Omega\ (y_{grnd})$, position in the image coordinate as a function of y_{grnd}, is used for the transformation from $h_{app}(y_{grnd})$ to $h_{true}(y_{grnd})$, calibration between w and y_{grnd} is required. We used cameras already in practical operations by officers in this paper. Therefore, geometrical condition of camera setup is not rigid, and it is difficult to perform calibration between w and y_{grnd} from the geometrical condition. In addition, since image resolution steeply declines along with the depth toward the vanishing point, it is difficult to perform the calibration accurately. Consequently, we decided to calibrate $\Omega\ (y_{grnd})$ from the traffic image in direct manner as described in the experimental results, Section 6.1.

$$P \times \Omega(w) = P + z = P + Z \times \frac{Q}{w + Q} \tag{3}$$

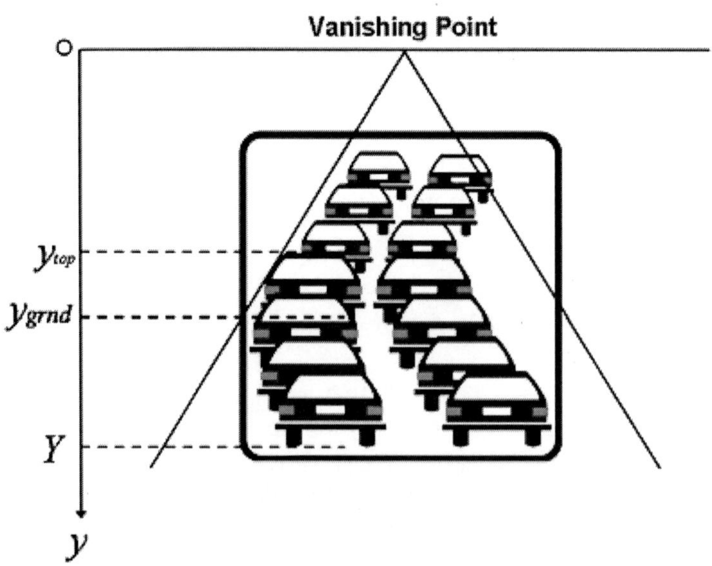

Figure 5. Definition of coordinate in the image coordinate

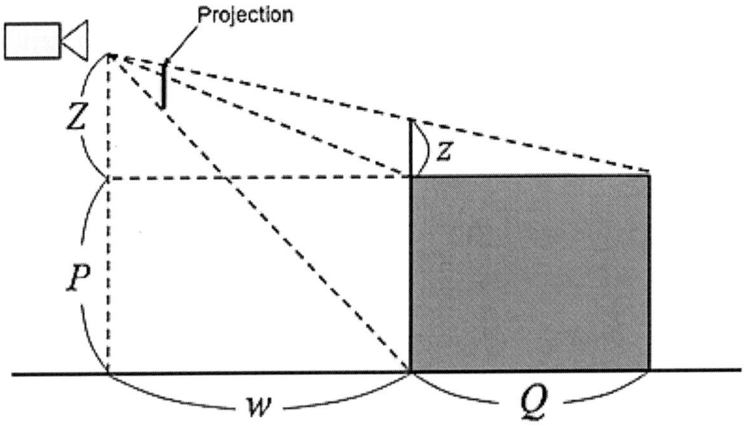

Figure 6. Side view of appearance height and true height of a vehicle projected onto the image coordinate

Equation 4 assumes that true height of the vehicle in the image coordinate is proportional to the distance from the vanishing point in the image coordinate. Consequently, Eqs. 5 and 6 are obtained from Eqs. 1 to 4. As same as Eq. 2, y_{grnd} is estimated from y_{grnd} itself in Eq. 6, and a feedback optimization process is necessary. Following descriptions are about the procedure to estimate y_{grnd} of the preceding vehicle occluded by the following vehicle.

$$h_{true}\left(y_{grnd}\right) = h_{true}\left(Y_{grnd}\right) \times \frac{y_{grnd}}{Y_{grnd}} \tag{4}$$

$$\begin{aligned} h_{app}\left(y_{grnd}\right) &= h_{true}\left(y_{grnd}\right) \times \Omega\left(y_{grnd}\right) \\ &= h_{true}\left(Y_{grnd}\right) \times \frac{y_{grnd}}{Y_{grnd}} \times \Omega\left(y_{grnd}\right) \end{aligned} \tag{5}$$

$$y_{grnd} = \left\{1 - \frac{h_{app}\left(Y_{grnd}\right)}{Y_{grnd}} \times \frac{\Omega\left(y_{grnd}\right)}{\Omega\left(Y_{grnd}\right)}\right\}^{-1} y_{top} \tag{6}$$

In the first step, $h_{app}(y_{grnd})$ of a vehicle is measured from the image, and the vehicle is classified into two categories of large vehicles as trucks and small vehicles as passenger cars. Vehicle regions are obtained from segmentation results from the S-T MRF model, and most of the vehicles are not occluded at the entrance area of the camera view. When the vehicle belongs to the large vehicle category, it is regarded as $\Omega\ (y_{grnd}) = 1$. Since large vehicles as trucks are higher than 2.5 m, the height of the camera position, roofs of the vehicles do not appear in the image. When the vehicle belongs to the small vehicle category, $\Omega\ (y_{ground})$ is obtained from the calibration results.

In the second step, y_{top} of the preceding vehicle is obtained by extracting the edge at the top of the vehicle region. The preceding vehicle and the following vehicle are segmented by the S-T MRF model against the occlusion at this moment, and the top edge of the preceding vehicle appears in the image while the bottom edge is occluded. Finally, y_{grnd} is obtained from Eq. 6 by substituting the value of y_{top}. However, right side of Eq. 6 has y_{grnd} as a valuable. Therefore, the iterative procedure should be performed as substituting y_{grnd} on the left side into y_{grnd} on the right side of Eq. 6 for the next loop of the iteration. In the first loop of the iteration, the value of y_{top} is substituted into y_{grnd} on the right side, iteration of single loop appeared as sufficient to pursue the precise result of y_{grnd} from our experiment.

Vehicle trajectory of each vehicle can be obtained by sequencing the grounding points of the vehicle in image coordinate, and the vehicle trajectory enables to estimate vehicle velocity. Thus, estimated vehicle velocity should be transformed into the velocity in the real world coordinate. Usually, calibration about the transforming parameters is performed based on solving camera parameters and setting condition. However, since the cameras used in this paper are already in practical operations aside heavy traffic, such the calibration is quite difficult. Therefore, we adopted more practical method for the calibration in this paper. Ultrasonic wave sensors are installed on the road side of our experimental field. In the highway of this paper, ultrasonic wave sensors are standardized as to show errors of less than 5% from the true value in the velocity measure. Calibrations between the image coordinates and the real world coordinates were performed to keep consistency of average velocities between the ultrasonic wave sensors and the image sensors.

5. IMPLEMENTATION OF SEMANTIC LAYER FOR INCIDENT DETECTION

Generally, it is quite difficult to detect the moment of the collision from an image. Therefore, algorithms to detect abnormally slowdown vehicles and stalling vehicles have been developed to detect incidents in which vehicles are stalling because of accidents as well as their own troubles. Slowdown vehicles and stalling vehicles are easily detected in sparse traffics. However, it is difficult to detect them from the congested or jammed traffics because a majority of vehicles move in similar velocities in congested or jammed traffics. In this paper, we designed semantic layer algorithms to classifying the behavior of each vehicle and to detect the incident by analyzing the behaviors of interacting vehicles.

5.1 Behavior-class Functions

From the previous section, we can obtain spatio-temporal trajectory of each vehicle. By classifying the trajectories, vehicle behaviors can be determined as slowdown, stalling, and lane changing. The following are examples of functions representing behaviors of vehicles such as slowdown, stalling, changing lanes, vacant space ahead of the vehicle.

- *typeOfSpTrajectory*(V.*i*, V_j,*t*): Returns 'Changing lanes' when vehicle V.*i* changes its lane to avoid vehicle V.*j*. Otherwise, returns 'Ordinary'. 'Changing lanes' includes that vehicles change its lane and has six types of traveling: 'left-to-right', 'right-to-left', 'center-to-left', 'left-to-center', 'center-to-right', and 'right-to-center'. The six types of 'Lane changing' are discriminated by evaluating similarities of a vehicle trajectory to the clustered trajectories of the particular types.

- typeOfSpTrajectory $(V.i, L_m^k, t)$: Returns 'Changing lanes' when vehicle V.*i* changes its lane to avoid the m–th discretized location along the k–th lane. Otherwise, returns 'Ordinary'.

- *typeOfTpTrajectory*(V.*i*, *t*): Returns 'True' when VehicleV.*i* is stalled at time t, and 'False' otherwise. The threshold between stalled or not is determined by statistical analysis of the vehicle behaviors.

- *isSpace*(V.*i, t*): Returns 'True' when there is the vacant space ahead of the VehicleV.*i* along the vehicle traveling lane at time t, and 'False' otherwise. It should be noted that there is no traffic flow in any lanes, this function returns 'False'.

5.2 Event-class Functions

Event-class functions are described by combining behavior functions to determine the types of events. The following are descriptions of logics for the incident detection.

Logic 1: If average velocity of the vehicles is less than TH_{ave_vel}, the traffic should be determined as congested, otherwise the traffic should be determined as free flow.

Logic 2: If a vehicle is stalling at the prohibited area in the period longer than TH_{stall_time}, the scene should be determined as an incident.

Logic 3: If there appeared the vacant space ahead of the stalling vehicle while the traffic is free flow, the scene should be determined as an incident.

Logic 4: If there occurred lane changing behaviors over the stalling vehicle more than $TH_{lchange}$ times, the scene should be determined as an incident.

Above algorithms for the incident detection were developed and proved as successful for the incident detection at relatively high angle in the open aired highway [26]. In this paper, the algorithms are applied to the tracking results by applying the interlayer feedback algorithm to low angle images in the tunnel. We adopted TH_{stall_time} = 10 sec, and $TH_{lchange}$ = 3 in the following experiments according to the publication [26], and 20km=hour is defined according to operational rules in the highway.

6. EXPERIMENTAL RESULTS

Experiments were performed by using video images recorded by the cameras at three different locations inside the tunnel. Those three locations consist of straight, separating, and merging traffic. In this section, experimental results of vehicle segmentations and incident detections are described.

6.1 Calibration of Ω (y_{grnd})

Figure 7 shows calibration results of Ω (y_{grnd}). The plots were obtained by measuring the ratio between true height and appearance height of vehicles directly from images. All the vehicles used for the calibration were not occluded by the following vehicles, in order to obtain y_{grnd} directly from the images. Each plot corresponds to each vehicle measured. For these plots, horizontal axis represents y_{grnd} of the vehicle, and vertical axis represents Ω (y_{grnd}). Diversity of Ω (y_{grnd}) is due to variations in vehicle types, shapes of rear and top of the vehicles. Consequently, Ω (y_{grnd}) was estimated as an average of the vehicles. In Fig. 7, smaller y_{grnd} means close to the vanishing point, and smaller Ω (y_{grnd}) means true height of the vehicle is close to appeared height.

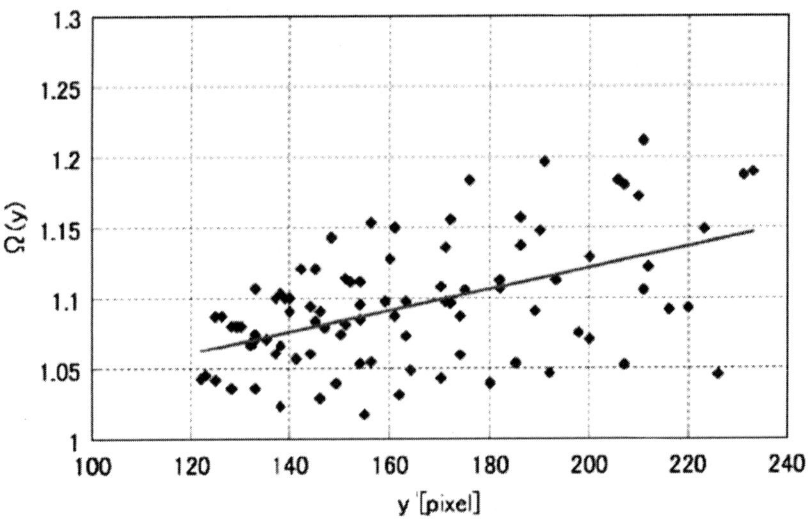

Figure 7. An example of Ω(y) calibration

6.2 Segmentation and Trajectory Estimation

Figure 8a and b exemplifies segmentation result by the S-T MRF model and the interlayer feedback algorithm. The S-T MRF model failed to segment vehicles traveling with the similar velocities and occluding along all the way in the view, as shown in Fig. 8a. The interlayer feedback algorithm succeeded to segment the two vehicles, and their trajectories were correctly obtained, as shown in Fig. 8b.

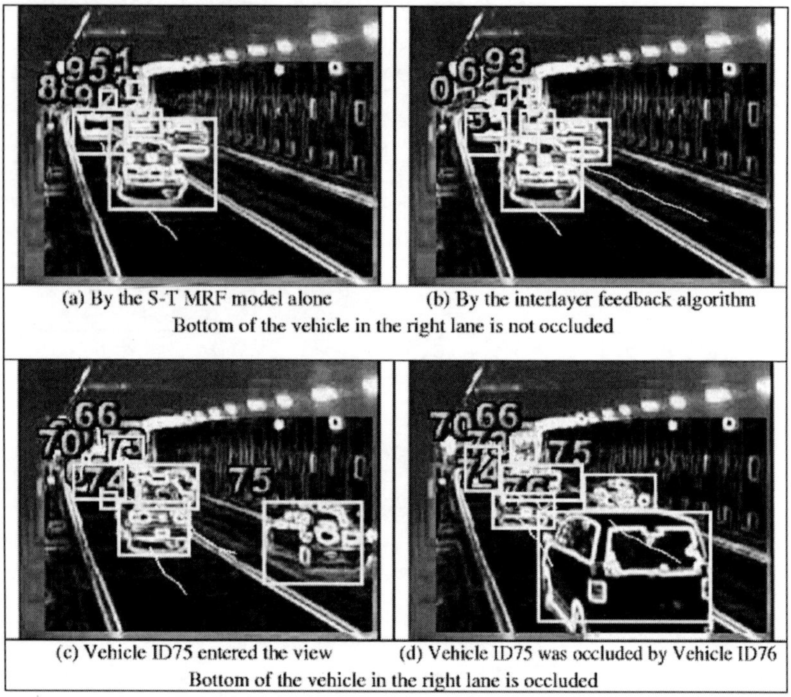

Figure 8. Segmentation results and trajectory estimations by the interlayer feedback process. **a** By the S-T MRF model alone. **b** By the interlayer feedback algorithm. Bottom of the vehicle in the right lane is not occluded.**c** Vehicle ID75 entered the view. **d** Vehicle ID75 was occluded by Vehicle ID76. Bottom of the vehicle in the right lane is occluded

Segmentation performance was evaluated by using 70-min video recorded by the three cameras in the tunnel. Totally, 8,563 vehicles were passed through the views of the three cameras. As a result, 6,577 segments were obtained by the S-T MRF model, and 8,964 segments were obtained by the interlayer feedback algorithm. Out of 8,864 segments by the interlayer algorithm, 7,856 segments were correctly appeared to include a single vehicle, and 107 segments appeared to include multiple vehicles. Nine hundred one segments appeared to be excessively divided segments. Consequently, in a rough description, contributions on vehicle segmentation against heavy occlusion would come from the S-T MRF model for about 77% and from the interlayer feedback for about 27%.

Figure 8c and d shows an example of trajectory estimations of the preceding vehicles over the occlusions by the following vehicles. Although the preceding vehicle of ID75 in the right lane was occluded by the following vehicle of ID76 in the left lane, the trajectory of the vehicle of ID75 was estimated precisely. Usually, the grounding position of the vehicle tail appears inside the bounding box of the vehicle segment. However, grounding position of the vehicle should be outside of the bounding box when the bottom part of the vehicle was occluded by the following vehicle. In Fig. 8d, each line tailing from the bottom of the vehicle connects the grounding positions y_{grnd} for the past five image frames of the vehicle. The latest grounding point of the vehicle of ID75 exists outside of the bounding box. Since the bottom part of the vehicle of ID75 was occluded by the vehicle of ID76, this result was successful.

6.3 Incident Detection Results

Figure 9 shows a result of stalling vehicle detection. Figure 9a shows the scene soon after the vehicle stalled at the road side. The algorithm determined that the vehicle is stalling in the prohibited area over the emergency lane. After the length of the vehicle stalling exceeded the threshold, the state of the algorithm turned into 'incident' as in Fig. 9b. Even if there was no emergency lane, the algorithm can conclude the scene as 'incident' by detecting stalling vehicle and passing vehicle in the same lane as in Fig. 10a and b.

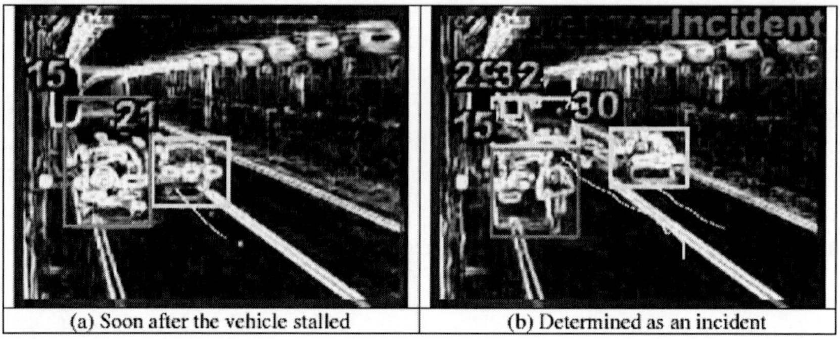

Figure 9. Detection of stalling vehicle. **a** Soon after the vehicle stalled. **b** Determined as an incident

Figure 10 shows image sequences of incident detection process by the S-T MRF model alone and the interlayer feedback algorithm. The incident could not be detected by the S-T MRF model alone, since vehicles could not be segmented because of heavy occlusions as shown in Fig. 10a. However, the incident was detected successfully by the interlayer feedback algorithm and edge pattern classification, since vehicles were segmented precisely against heavy occlusions as shown in Fig. 10b. In this case, the algorithm detected the incident by detecting frequent lane changing behaviors against a stalling vehicle. Such the behavior analyses over the congested traffic were enabled by the vehicle segmentation using the interlayer algorithm and the trajectory estimations against occlusions.

Consequently, experimental results were summarized as shown in Table 1, in terms of success rates of incident detection. For the experiments, 32 incidents were edited from the 6 month video stream of the cameras. Videos of 16 incidents were used as learning data to determine thresholds to detect slowdowns, stalling, and the frequent lane changing in the same location to avoid the incident vehicles. The other videos of 16 incidents were used as test data to examine the performance of the algorithm. Success rates for incident detections were improved by the interlayer feedback algorithm to 88% for test data and 91% in total of learning data and test data, whereas they were 44% and 53% by the S-T MRF model alone, respectively.

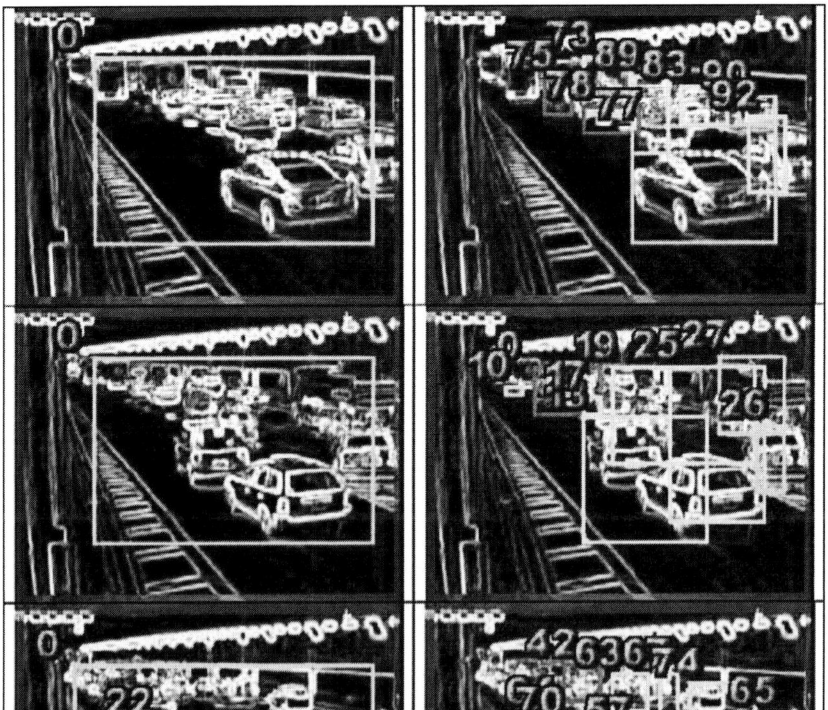

(a) By the S-T MRF model | (b) By the interlayer feedback algorithm

Figure 10. A result of Incident detection: lane changing around the stalling vehicle

Table 1. Incident detection results: success rates

	Learning Data	Test Data	Total
By the S-T MRF model alone	10/16=63%	7/16=44%	17/32=53%
By the interlayer feedback	15/16=94%	14/16=88%	29/32=91%

7. CONCLUSIONS

In this paper, an algorithm for incident detection in the tunnel was developed. We developed the interlayer algorithm to combine the S-T MRF model and the algorithm in the Morphology layer, in order to resolve heavy occlusion problems in heavy traffic with the low angle image. The algorithm was proved to be effective for the segmentation of vehicles in the heavy occlusion images though the examinations using heavy occlusion images. Consequently, the algorithm could obtained precise trajectories of the vehicles, and contributed to the incident detections from the congested traffics in tunnels.

Since the incidents in tunnels may cause additional accidents or fires that may result in fatal disaster, it is important to detect them as soon as possible. The disaster would be rather serious in the heavy traffics of urban tunnels. The technology in this paper is expected to enhance the safety of tunnel traffics and the efficiency of tunnel operations.

REFERENCES

1. Kolling, H., Nagel, H.: 3D pose estimation by directly matching polyhedral models to gray value gradients. IJCV **23**(3), 283–302 (1997)

2. Peterfreund, N.: Robust tracking of position and velocity with Kalman Snakes. IEEE Trans. Pattern Analysis and Machine Intelligence (PAMI), vol. 21, no. 6, pp. 321–331, 1988

3. Kass, M., Witkin, A., Terzopoulos, D.: Snakes: active contour models. Int. J. Comput. Vis. **1**, 564–569 (1999)

4. Smith, S.M., Brady, J.M.: ASSET-2: real-time motion segmentation and shape tracking. IEEE Trans. Pattern Analysis and Machine Intelligence (PAMI), vol. 17, no. 8, pp. 814–820, 1995

5. Stauffer, C., Grimson, W.E.L.: Adaptive background mixture models for real-time tracking. Proc. of CVPR, 1999, vol. II, pp. 246–252, Jun 1999

6. Leuck, H., Nagel, H.: Automatic differentiation facilitates of-integration into steering-angle-based road vehicle tracking. IEEE CVPR'99, pp. 360–365

7. Gardner, W.F., Lawton, D.T.: Interactive model-based vehicle tracking. IEEE Trans. PAMI, vol. 18, no. 11, pp. 1115–1121, 1996

8. Weber, K., Malik, J.: Robust multiple car tracking with occlusion reasoning. Proc. of ECCV'94, Vol. I, Lecture Notes in Computer Science 800, pp. 189–196. Springer-Verlag

9. Hsieh, J.-W., Yu, S.-H., Chen, Y.-S., Hu, W.-F.: Automatic traffic surveillance system for vehicle tracking and classification. IEEE Transactions On Intelligent Transportation Systems, vol. 7, no. 2, pp. 175–187, June 2006

10. Kanhere, N.K., Birchfield, S.T.: Real-time incremental segmentation and tracking of vehicles at low camera angles using stable features. IEEE Transactions On Intelligent Transportation Systems, vol. 9, no. 1, pp. 148–160, March 2008

11. Wang, C.-C.R., Lien, J.-J.J.: Automatic vehicle detection using local features - a statistical approach. IEEE Transactions On Intelligent Transportation Systems, vol. 9, no. 1, pp. 83–96, March 2008

12. Zhang, W., Wu, Q.M.J., Yang, X., Fang, X.: Multilever framework to detect and handle vehicle occlusion. IEEE Transactions on Intelligent Transportation Systems, vol. 9, no. 1, pp. 161–174, March 2008

13. Ambai, M., Ozawa, S.: Vision-based traffic monitoring system for various traffic situations using monocular camera. 13th World Congress on ITS, London, October 2006

14. Cucchara, R., Piccardi, M., Mello, P.: Image analysis and rule-base reasoning for a traffic monitoring system. IEEE/IEEJ/JSAI ITS conf., Tokyo, pp. 758–763, Oct. 1999

15. Jung, Y., Lee, K., Ho, Y.: Content-based event retrieval using semantic scene interpretation for automated traffic surveillance. IEEE Trans. on ITS, Sep. 2001

16. Zhang, K., Taylor, M.: A new method for incident detection on urban arterial roads. 11thWorld Congresson ITS, Nagoya, Oct 2004

17. Srinivasan, D., Jin, X., Cheu, R.L.: Evaluation of adaptive neural network models for freeway incident detection. IEEE Trans. ITS, vol. 5, no. 1, March 2004

18. Byun, S.C., Choi, D.B., Ahn, B.H., Ko, H.: Traffic incident detection using evidential reasoning based data fusion. Proceedings of ITS 1999, 6th World Congress on ITS

19. Thomas, T., van Berkum, E.C.: Detection of incidents and events in urban networks. IET Intell. Transp. Syst. **3**(2), 198–205 (2009)

20. Zheng, P., McDonald, M., Jeffery, D.: Event detection based on loop and journey time data. IET Intell. Transp. Syst. **2**(2), 113–119 (2008)

21. Shehata, M.S., Cai, J., Badawy, W.M., Burr, T.W., Pervez, M.S., Johannesson, R.J., Radmanesh, A.: Video-based automatic incident detection for smart roads: the outdoor environmental challenges regarding false alarms. IEEE Transactions on Intelligent Transportation Systems, vol. 9, no. 2, pp. 349–359, June 2008

22. Versavel, J.: Improving road and tunnel safety via incident management: implementing a video image processing system. 16th World Congress on ITS, Stockholm, September 2009

23. Austin, M., Brock, M., Carter, E., Metcalfe, F., Orme, L.: Implementation and evaluation of an adaptive CCTV display system for incident detection for the highways agency. 16th World Congress on ITS, Stockholm, September 2009

24. Kamijo, S., Matsushita, Y., Ikeuchi, K., Sakauchi, M.: Traffic monitoring and accident detection at intersections. IEEE trans. ITS, vol. 1, no. 2, pp. 108–118, June 2000

25. Kamijo, S., Nishida, T., Sakauchi, M.: Occlusion robust and illumination invariant vehicle tracking for acquiring detailed statistics from traffic

images. To appear in IEICE Transactions on Information and Systems, vol. E85-D, no. 11, 2002

26. Kamijo, S., Harada, M., Sakauchi, M.: An incident detection system based on semantic hierarchy. Proceedings of ITSC 2004, pp. 853–858, Oct. 2004

Chapter 9

Wireless Sensor Networks for Condition Monitoring in the Railway Industry: A Survey

Victoria J. Hodge, Simon O'Keefe, Michael Weeks, and Anthony Moulds

ABSTRACT

In recent years, the range of sensing technologies has expanded rapidly, whereas sensor devices have become cheaper. This has led to a rapid expansion in condition monitoring of systems, structures, vehicles, and machinery using sensors. Key factors are the recent advances in networking technologies such as wireless communication and mobile ad hoc networking coupled with the technology to integrate devices. Wireless sensor networks (WSNs) can be used for monitoring the railway infrastructure such as bridges, rail tracks, track beds, and track equipment along with vehicle health monitoring such as chassis, bogies, wheels, and wagons. Condition monitoring reduces human inspection requirements through automated monitoring, reduces maintenance through detecting faults before they escalate, and improves safety and reliability. This is vital for the development, upgrading, and expansion of railway networks. This paper surveys these wireless sensors network technology for monitoring in the railway industry for analyzing systems, structures, vehicles, and machinery. This paper focuses on practical engineering solutions, principally, which sensor devices are used and what they are used for; and the identification of sensor configurations and network topologies. It identifies their respective motivations and distinguishes their advantages and disadvantages in a comparative review.

1. INTRODUCTION

Experts estimate that the railway industry will receive US$300 billion worth of global investment for development, upgrading, and expansion over the five years from 2009[43]. Ollier [98] noted that effective management of rail infrastructure will be vital to this development, upgrading, and expansion, particularly if coupled with a move to intelligent infrastructure [39]. A key part of the management will be condition monitoring. Condition monitoring detects and identifies deterioration in structures and infrastructure before the deterioration causes a failure or prevents rail operations. In simple condition monitoring, sensors monitor the condition of a structure or machinery. If the sensor readings reach a predetermined limit or fault condition, then an alarm is activated. However, this simplistic approach may lead to a large number of false alarms and missed failures [36]. It only provides local analysis but does not take advantage of the superior capabilities when the sensors are networked and their data processed collectively. Integrated data processing allows an overall picture of an asset's condition to be achieved and overall condition trends to be determined [97].

In recent years, networking technologies such as wireless communication and mobile*ad hoc* networking coupled with the technology to integrate devices have rapidly developed. The new technologies allow vast numbers of distributed sensors to be networked[5], [6], [37], [45], [122] to constantly monitor machines, systems, and environments. Wireless sensor networks (WSNs)[5], [134] are wireless networks of spatially distributed and autonomous devices. They use sensors to cooperatively monitor infrastructure, structures, and machinery. A typical WSN for railway applications is shown in Fig. 1. Each sensor node generally has a radio transceiver, a small microcontroller, and an energy source, usually a battery (see Section II-C for more detail). WSNs and data analytics allow the railways to turn data into intelligence [43]. They provide decision support through continuous real-time data capture and analysis to identify faults [52]. The data from distributed systems such as sensor networks are constantly monitored using classification [56], [57], prediction [85], or anomaly detection [61] to determine the current and future status of the distributed network. Lopez-Higuera *et al.* [78] developed a staircase of structural health monitoring, where the higher the level, the higher the complexity and functionality. The simplest Level 1 systems detect the presence of damage without locating it, whereas Level 2 provides location information. A Level 3 system is able to grade the degree of damage and a Level 4 system can estimate the consequences of the damage and remaining service life. Finally, a Level 5 system will comprise complex hardware, custom algorithms and software to allow the diagnosis and/or the prognosis and even to recommend the solution to a problem.

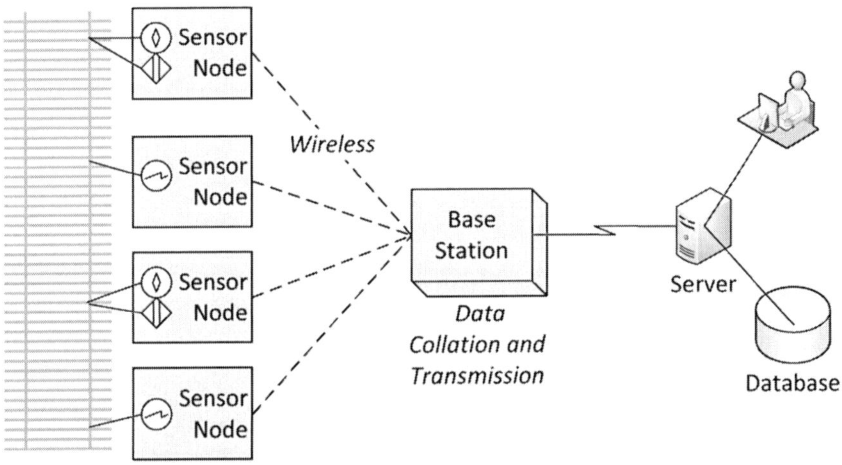

Figure 1. Figure shows a typical WSN setup for railway condition monitoring. Sensor devices are mounted on boards attached to the object being monitored; examples include track, bridges, or train mechanics. One or more sensors are mounted on a sensor board (node) (see also Fig. 2). The sensor nodes communicate with the base station using a wireless transmission protocol; examples include Bluetooth and Wi-Fi. The base station collates data and transmits it to the control center server possibly through satellite or GPRS. There are variations on this setup. In some systems, the sensor nodes may communicate directly with the server rather than via the base station. In other systems, the user accesses the data directly via the base station.

WSN monitoring provides continuous and near real-time data acquisition and autonomous data acquisition (no supervision is required); increased frequency of monitoring compared with manual inspection; improved data accessibility, data management, and data use compared with non-networked systems as all data can be collected and processed centrally; the ability to combine data from a wide variety of sensors; intelligent analysis of data to "predict and prevent" events using intelligent algorithms; the ability to turn data into information about the status of important structures, infrastructure and machinery; and, a global data view that allows trending information to be determined where degradation is happening slowly over a relatively long period of time.

WSN monitoring can be used to:

1 maintain process tolerances;

2 verify and protect machine, systems and process stability;

3 detect maintenance requirements;

4 minimize downtime;

5 prevent failures and save businesses money and time;

6 request maintenance based on the prediction of failure rather than maintenance running to a standard schedule or being requested following an actual failure.

There are a number of challenges with WSNs. They generate large amounts of data at rapid rates and often on an *ad hoc* basis. Data may be produced from multisources that have to be fused. The systems and structures monitored using sensors often exhibit complex behavior, which is difficult to understand and interpret [102]. Hence, the data must be carefully managed to provide a view of the system status. Sensor data are very noisy and sensors themselves can become defective wherever they are installed. Sensor data may contain errors, particularly where the sensors are subject to harsh conditions as this exacerbates sensor and communication failures. Sensor networks often have to be installed in challenging environments to be able to monitor structures and infrastructure. For example, Palo [99] noted that their system had to work in extreme conditions in Sweden, with a temperature range between and and with large quantities of snow. Grudén *et al.* [55] mounted sensors on the train's bogies to monitor bogie temperatures and noted that train environments are very harsh environments for electronics with high accelerations and large shocks. The sensors need to be carefully located to ensure their measurements are useful and do not replicate the measurements of other sensors, which can skew the distribution of the collected data. The type of sensor used needs to be carefully considered to ensure the maximum value and the best quality data. WSNs can use a set of homogeneous or heterogeneous sensors. Sensors are often located away from energy supplies, thus require either batteries or some form of local energy generation to power them. If there are errors in transmission across the WSN, then data may be missing. These last two points form a paradox, WSNs need to minimize energy usage yet communication needs to be maximally efficient and communication requires energy.

This survey paper describes WSNs for railway condition monitoring focusing on systems described in the academic literature. In this survey, "sensor" refers to an individual device such as an accelerometer or strain gauge. "Node" refers to a whole sensor unit that comprises the sensor, a power supply, a data transmitter/receiver, and a microcontroller.Section II of the survey discusses the design and range of sensor devices, Section III analyzes the network topology and transmission techniques, andSection IV reviews the sensors used in the literature to monitor railway equipment and structures.Section V draws conclusions, and Section VI examines future work in the area of railway monitoring.

2. SENSOR DESIGN

There are a multitude of sensors types used in railway condition monitoring for analyzing different aspects of structures, infrastructure, and machinery (see Table I in the Appendix). The devices in Table I are transducers. Transducers convert energy from one form to another. In the case of sensors, the device typically converts a measured mechanical signal into an electrical signal. Most railway sensors fall under the umbrella type microelectromechanical systems (MEMS). MEMS are small, integrated devices, or systems that combine electrical and mechanical components.They are cheap and efficient[70]. Sensor design requires a tradeoff between functionality and power consumption, with functionality often coming at the cost of power. Condition monitoring systems for railways are often deployed in remote or inaccessible locations, where there is no wired power supply available. Hence, the sensors must receive power from either batteries or local energy generation. MEMS sensors have the advantage that they can be produced to consume ultralow power[19].

2.1 Measurement Sensors

Table I provides an overview of the sensor types used in railway condition monitoring and the measurements produced.

Sensor devices may also return additional ambient measurements such as temperature and humidity, and sensor data can be combined with the measurements from other systems. Rabatel *et al.*[103] noted that the behavior of assets is often affected by external factors. These external factors are encapsulated by contextual data, which describe the ambient running conditions. It is important to take them into account and build models for condition monitoring that consider these ambient conditions. Hence, Elia *et al.*[44], Rabatel *et al.* [103], and others incorporated wider ranging contextual data with the sensor data. Elia *et al.*[44] included train speed, position along the line (longitude and latitude from GPS system), and wind speed and direction when analyzing bogie vibrations, whereas Rabatel *et al.*[103] included structural criteria (route) and environmental criteria (weather, travel characteristics, travel duration) when analyzing bogie temperatures.

2.2. Sensor Nodes

Sensor devices are mounted on boards. The boards form a platform combining mobile computing and wireless communication (media access control (MAC), routing and error detection) with sensor devices as shown in Fig. 2. Many sensor nodes are moving toward enabling data interrogation and autonomous data processing to identify anomalies *in situ*. Hence, many data processing algorithms have also been embedded in wireless sensor nodes for autonomous execution [80]. The boards generally comprise one or more wireless sensors, a microcontroller, transceiver, data storage (memory), and a power source. If the data sampling rate is low, for example, when monitoring the growth of a crack over a period of time, then the microcontroller, memory, and transceiver power can all be low. However, for frequent sampling such as monitoring the temperature of a mechanical bearing every second, then the microcontroller, memory, and transceiver must be more powerful. Lynch and Loh[80] provided a thorough overview of sensor boards for WSNs.

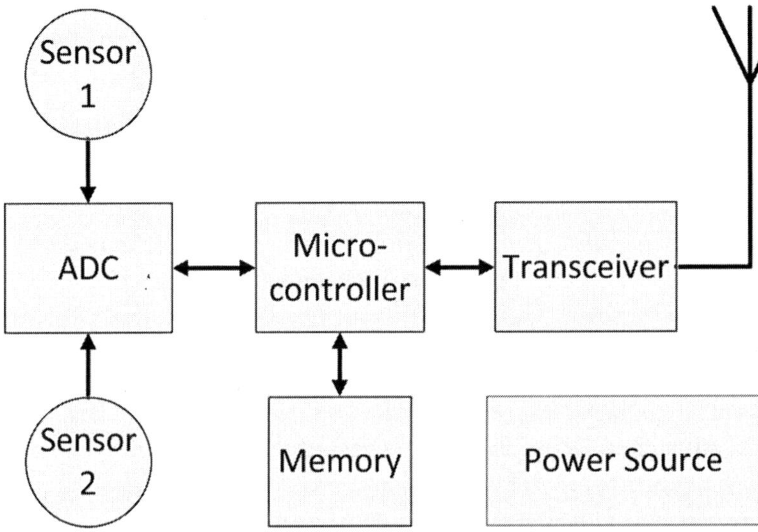

Figure 2. Diagram of the composition of a typical sensor node. Multiple sensors may be mounted on the node. The microcontrollers used in sensor nodes are ultralow-power microcontrollers to conserve energy.

2.3. Sensor Power

Sensor nodes often use batteries as their energy source. Batteries have a finite life. The batteries in an accelerometer monitoring a train pantograph only lasted 20 days [97].Chebrolu *et al.* [32] were able to extend battery life to 1.5 years for 4 AA batteries by only waking their sensors when required and sleeping otherwise. However, even with extended life, fitting and replacing batteries is not always possible in remote and inaccessible railway locations. The nodes may also be embedded in the monitored structure or mechanics of a train, where no access to batteries or wired connection is possible. Hence, authors have considered alternative energy sources for nodes. These need to be reliable to allow the WSN to function for long periods of time with no human involvement and able to generate sufficient power.

Nagode *et al.* [89], Nelson *et al.* [92], Nuffer and Bein [96], Pourghodrat [101], Tan *et al.* [120], and Wischke *et al.* [130] investigated ambient energy harvesting technology for powering WSNs. Piezoelectric materials can produce electricity as a response to mechanical strain such as the strain exerted on the track or sleepers by a passing train.Piezoelectric materials are robust, able to withstand harsh environmental conditions, and can deliver a large amount of energy[55]. Nelson *et al.* [92], Nuffer and Bein [96], Pourghodrat [101], Tan *et al.* [120], and Wischke *et al.* [130] investigated piezoelectric vibration harvesters that derive energy from the vertical displacement or vibration of tracks, ties, and sleepers. Wischke *et al.* [130] were able to demonstrate that sufficient energy could be harvested to supply a microcontroller with a radio-frequency (RF) interface. Tan *et al.* [120] noted that their rate of energy harvesting (1 mW) was lower than the typical rate of energy consumption of sensor systems (10–100 mW). Hence, they stored the harvested energy using supercapacitors and the sensors only drew energy when sufficient stored energy was available. Nagode *et al.* [89], Nelson *et al* .[92], and Tan *et al.* [120] all used inductive coils for harvesting. Electromagnetic harvesters such as induction coils have a higher power density than piezoelectric [69]. Nelson *et al.* and Tan *et al.* used the vertical movement of the track to generate energy. Nagode *et al.* designed and built two prototypes of a vibration-based electromagnetic (inductive coil) system. They both harvested mechanical energy from the movement of a rail wagon's suspension springs to power sensors mounted on the wagons. The first prototype used the movement of the springs to generate linear motion in magnets to produce a voltage. The second prototype converted the linear motion of the wagon's suspension coils into rotary motion then magnified and rectified the motion to turn a generator. Pourghodrat[101] developed a hydraulic energy harvester, which generated energy from compression and relaxation of fluid driving a hydraulic motor. It used the vertical displacement of the track for energy generation. Hydraulic harvesters can generate energy from smaller track deflections compared with electromagnetic, but the devices are still in development[101].

A number of authors have investigated using solar energy harvesting to power sensors.Bischoff *et al.*[19] investigated a solar energy rechargeable battery-powered base station. Sekula and Kolakowski[111] powered their piezoelectric strain sensors using accumulators permanently charged by photovoltaic modules. Solar power is a proven way to generate energy. However, it has a number of practical drawbacks for WSNs.Solar panels used to generate energy need frequent cleaning for optimal performance[55]. Sunlight varies on a daily basis, and sunlight is very limited in winter months in many global regions and power generation is adversely affected by bad weather. Hence, Wolfs *et al.*[131] powered accelerometers and gyroscopes using solar power with battery backup and Waki[125] powered a temperature sensor by a solar cell with a double-layer capacitor power supply for back up. These double layer capacitors have a longer life than a battery.

If the sensor nodes can be put into "sleep" mode and only woken when they are required, then their energy usage can be minimized. A number of bridge monitoring systems use this wake-on-event two-layer event detection. It requires two types of sensors: one set have lower power consumption to perform event detection and wake the second set of sensors using interrupts,[19], [32], [46], [70], [73] and [107]. Bischoff*et al.* [19], Feltrin [46] and Kruger *et al.* [70] detected approaching trains using ultralow-power MEMS acceleration sensors. These measured the vibrations of the bridge and detected the increased vibrations when a train approached. Similarly, Ledeczi *et al.*[73] used standard strain gauges to sense the approaching train and woke the system by detecting changes in stress on the bridge. Sala *et al.*[107] used activation sensors type not specified) to wake their system for monitoring steel truss bridges. In contrast, Townsend and Arms [123] used a similar technique but with only one set of track-mounted strain gauges, which varied their sampling rate depending whether a train is present. They operated in low sampling mode of approximately 6 Hz to detect the increase in strain when a train approached. Once detected, the sensors increased their sampling rate and generated a strain waveform for data analysis. The authors stated that this reduced the power from 30 mA for continuous mode to less than 1 mA for event detection. Chen *et al.*[35] introduced a prototype system that developed this further by detecting approaching trains, waking the measuring sensors on the bridge trusses and then downloading the measured data onto the passing train, which acts as a data mule. Another data muling approach is the BriMon system [32], where trains were fitted with a beacon. A bridge-mounted high-gain antenna detected the train's beacon 30+ seconds before the train approached and woke the battery-operated accelerometers on the bridge. The accelerometers measured the vibrations of the passing train on the bridge and uploaded their data to the passing train. This data muling only requires local data transmission, which reduces the power usage further.

3. NETWORK DESIGN

WSNs enable continuous real-time capture of data. However, WSNs need to be able to handle the harshness of outdoor long-term condition monitoring; often in hostile environments and must minimize energy usage as the nodes are not attached to a wired power supply. They typically use low-power sensors powered by batteries although authors are investigating alternative power supplies such as local energy generation. Hence, the network to enable data capture has to be carefully designed to overcome these factors and prevent transmission errors, latency, network outages, missing data, or corrupted data.

3.1. Base Station

The base station controls the sensor nodes and acts as a gateway for data transmission to a remote server. The sensor nodes use short-range communication such as Wi-Fi or Bluetooth to transmit data to the base station. The base station uses long-range communication such as GPRS or satellite to transmit collated data back to a server at a control center. It has a more powerful processor and more memory than the sensor nodes to allow it to collate data from multiple sensors. Hence, the base station requires more power. For example, train monitoring systems have battery-powered sensor nodes in the carriages but place the base station in the locomotive, where more power is available from the train engine.

3.2. Relay Nodes

Sensor nodes are energy-constrained, thus only having a short transmission range. If they cannot reach the base station, then higher capability relay nodes may be used. This is a particular issue in restricted environments such as railway tunnels, along the length of trains or along railway tracks. Relays ensure connectivity by relaying the data signal from the sensor node to the base station via one or more relay nodes using wireless multihop paths. Network nodes can be multifunctional. Hoult *et al.*[62] and Shafiullah*et al.* [112] both used sensor nodes that also function as relay nodes to transmit data.

3.3. Network Model

WSNs for railway applications follow the open systems interconnection model [137]. A typical WSN has five layers in a protocol stack with three planes [5], [6] to transmit data from the sensor nodes to the base station. The five layers from lowest to highest are the **physical layer** defines how the sensors transmit their data to the network. The**data link layer** specifies the network topology and connects nodes to each other. It uses MAC to control data access and permissions and a logical link control to control error checking and data packet synchronization. This is a key layer in railway monitoring WSNs as they may be monitoring difficult environments such as tunnels or monitoring moving trains. The topology must be carefully designed. The **network layer** routes the data through the network as packets. Railway WSNs are energy constrained; thus, routing has to be carefully designed to work within the available energy supply. The **transport layer** controls the sending and receiving of data. Finally, the **application layer** allows application software to access the data.

The size of the transmission varies from bits in the lowest (**physical**) layer, bytes, and frames in the **data link** layer, packets in the **network** layer, segments in the**transport** layer, and data in the highest (**application**) layer.

The **power management plane** manages the power consumption of sensor nodes. The **mobility management plane** identifies and records the movement of sensor nodes to ensure that data can be routed from the node to the control center and so that sensor nodes are aware of their neighbors for communication. This is important for WSNs monitoring moving trains. The **task management plane** balances and coordinates network regions (groups of nodes) and ensures that they generate data according to their power level.

3.4. Sensor Network Topology (Data Linkage)

The topology is constrained by the requirements of the monitoring and by the physical environment. Sensor nodes can be arranged in either an *ad hoc* or a preplanned configuration [134]. In *ad hoc* arrangement, sensor nodes are randomly placed in the monitored area. In preplanned arrangement, sensor nodes are arranged in either a grid, optimal placement [100] (described next), 2-D or 3-D placement [134].

Determining the optimum node placement is a complex task and often requires a tradeoff. The network configuration can be optimized against a number of different constraints. A network may minimize relay nodes, may need to ensure a minimum level of service (include a certain level of redundancy), minimize energy usage to preserve battery life, or may need to ensure accessibility of the nodes[116]. For example, Pascaleet *al.* [100] optimized a WSN for signaling and control in railway stations to calculate the minimum number of nodes required. The authors had to optimize the tradeoff between reliability versus communication delay. Determining the best settings proved difficult as it required iterative graph optimization, which is computationally complex. They split the optimization into two tasks: deployment optimization to minimize the number of nodes while maintaining a minimum level of reliability, and scheduling optimization which allocates nodes while maintaining a maximum level of delay. The authors concede that the reliability versus communication delay tradeoff cannot be solved for all situations. The railway environment can also place constraints on the network topology. Signal propagation within a rail tunnel is different from propagation outside[77], [88], [116]. The complex geometry and rough surfaces of rail tunnels degrade RF signals. Conversely, the regular, tube-like shape can act as an "oversized waveguide"[88]. Therefore, Stajano *et al.* [116] and Liu *et al.* [77] developed models to predict the signal strength at each node in a tunnel. Node placement can then be optimized using an algorithm based on parameters such as transmission strength, battery life, and redundancy between the sensor nodes and the base station.

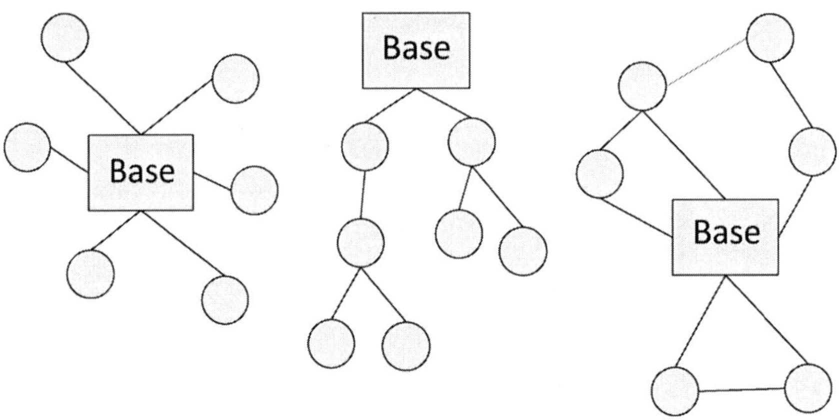

Figure 3. Diagram of a star, tree, and mesh network topology (from left to right) where the circles represent sensor nodes and the boxes are base stations.

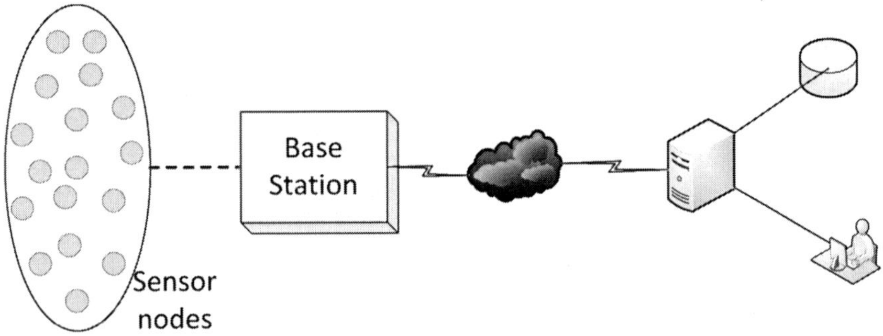

Fig. 4. In a typical WSN, the sensor nodes are arranged in a star, tree, or mesh topology and transmit data to the base station, which relays the data to the control center via longer range communications.

The natural topology choices available for WSNs in rail applications are star, tree, and mesh topologies (see Figs. 3 and 4). In a star topology (single point-to-multipoint), there is a single base station that can send and/or receive messages to/from a number of remote sensor nodes. The remote nodes can only communicate directly with the base station but not with each other. This topology is simple; the point-to-point connection will result in fewer transmissions and fewer collisions; and, it has low communication latency. The disadvantages are the base station must be within transmission range of all nodes, remote nodes must increase their transmission power to overcome signal attenuation and loss, and it is not robust due to its dependence on a single base station to manage the network.

A tree topology is a hierarchy of nodes, with the root node serving client nodes that, in turn, serve other lower level nodes. In a tree, the nodes can be grouped at each level. The network is scalable. Messages pass from the sensor nodes through the tree branches to the root. Sensor nodes can communicate with their parent or with other nodes in a group within transmission range.

In contrast, in a mesh topology, any node in the network can communicate with any other node in the network that is within transmission range. This enables relaying of messages (multihop communication) keeping the transmission power low, saving energy compared with transmitting directly from sensor nodes to base station in a single hop, and allowing nodes to communicate indirectly with nodes outside their transmission range. The network can be easily extended by adding new nodes and also incorporates fault tolerance as failed nodes can be bypassed using alternative routes.

3.5. Communications Medium

There are many communication techniques used in WSNs in railways [3], [108]. Intersensor communications and sensor to base station transmission are usually short range. The base station transmits gathered data back to the control center, and this requires long-range communication. WSNs can use technologies based on standard mobile telephony (Bluetooth, GSM, GPRS, and UMTS) or broadband techniques such as Wi-Fi (IEEE 802.11, 2007), wireless personal area networks (WPANs) (IEEE 802.15.4 and ZigBee[11]) or WiMax [2] (IEEE 802.16.2, 2004). ZigBee enhances IEEE 802.15.4 by including authentication, data encryption for security, and data routing and forwarding. Mobile telephony and WiFi are peer-to-peer only, whereas WPANs and WiMax can connect many devices in a mesh topology, where any node can communicate with any other node. In addition, Wi-Fi, WPANs and WiMax allow communication at much higher speeds and bandwidths than mobile telephony and are cheaper to set up. However, mobile telephony has better coverage and range than Wi-Fi and WPANs. WiMax can operate over long distances. However, there is a tradeoff: either high bitrates or long distances but not both. An important recent communication advance has been GSM-R (Global System for Mobile Communications-Railway), which is an adaptation of GSM telephony for railway applications. It is designed for information exchange between trains and control centers. Its main advantages are its low cost and worldwide availability. Another common technique for railways is the radio system Terrestrial Trunked Radio, which is a standard for data communication for closed user groups[108]. It is a private mobile telephone system and is easily deployed using a series of antennas at stations or control centers along the railway routes.

3.6. Transmission and Routing

Sensor nodes use the greatest amount of energy during data communication. The transceiver in the sensor node contains frequency synthesizers, mixers, voltage control oscillators, phase-locked loops, and power amplifiers[6]. These all consume power. Thus, data transmission and routing has to be carefully designed to minimize the energy usage.

When messages are transmitted through tree or mesh networks, they have to be routed to ensure they pass from source to destination (sensor node to base station). Message latency increases as the number of hops required to transmit the message increases; thus, the routing protocol must be carefully designed for the particular WSN topology.A number of routing protocols have been proposed recently (see[4]). If the base station fails in any topology, then the routing also fails. Thus, when Mahlknecht and Madani[81] used a daisy chain (tree) topology, they ensured that the nodes in the chain could also directly communicate with the global communication network if no base station was accessible. Another important consideration of multihop communication is interference prevention. For train-mounted WSNs, if two trains, Train A and Train B, are within radio communications range of each other, then data from sensors on Train A must never be transmitted to Train B and *vice versa*[104].Reason *et al.* [104] used identifiers to bind the sensors to the individual train's network.

The sensor nodes need to transmit data reliably via the network despite having only a limited power supply available. Hence, energy efficiency and reliability are both vital for WSNs. One important disadvantage of multihop (tree or mesh), in general, is that the power consumption of the nodes near the base station is much higher as they are used in data transmission to the base station and hence suffer energy drain. A possible solution to minimize energy usage and maximize fault tolerance is to use multihoming. This is used in the Internet and provides multiple connections (multiple routes) between points thus introducing a degree of fault tolerance and ensuring that no extra energy is required to transmit data. Aboelela *et al.*[1] mapped every node in the network to two clusters. One set of nodes operated as guardians of the clusters and cluster nodes transmitted via two designated cluster guardians. Each cluster guardian thus acted as a relay node and forwarded data from one cluster to the other or to the base station for processing.

Another issue with the *ad hoc* nature of WSNs is transmission interference. The MAC protocol must minimize both power usage and data broadcasting collisions. WSNs use IEEE 802.15.4, which uses carrier sense multiple access with collision avoidance multiplexing. This allows multiple nodes to access the same channel at different times without interference. However, the 2.4-GHz ISM band used by IEEE 802.15.4 is crowded causing variable and unpredictable transmission. Therefore, Liu *et al.*[76] investigated frequency diversity (FD) using modified sensors to monitor an underground railway tunnel where the confined environment made data transmission difficult. In FD, the data are transmitted using several frequency channels. Liu *et al.*[76] found that applying FD improved transmission for WSNs in confined environments such as tunnels. Shafiullah *et al.* combined energy efficient MAC protocols[114] with multifunctional nodes [112]. They developed energy-efficient protocols for IEEE 802.15.4/ZigBee multihop (daisy chain) networks mounted in train carriages. The nodes acted as both sensors and routers to both take measurements and forward data to a base station in the train locomotive. One MAC protocol (E-BMA) is optimal for low-medium traffic and the second (EA-TDMA) is optimal for medium-high traffic. In E-BMA, sensor nodes use piggybacking to reserve transmission slots rather than sending control messages to reduce transmission and, thus, reduce energy usage. E-BMA also waits for one extra time step to determine if there are multiple packets to transmit and sends them together. EA-TDMA divides the transmission channel into time slots and allocates slots to sensor nodes to avoid collisions. The nodes wake up in their slot, transmit if they have data or switch off immediately if they have no data, and sleep during the remaining time slots to conserve energy.

Salaberria *et al.* [108] developed an alternative multitransmissiontechnique using GPRS. They used two phonecards belonging to different telephone providers on board eachtrain to increase the network coverage. The hardware can then switch between networks depending on network availability.

4. MONITORING SYSTEM

The data from sensors is treated as either a time series, where data are produced continuously or periodically, or a sequence of readings where data is generated *ad hoc*, for example, generated every time a train passes. The data can be monitored by searching for thresholds (triggers), known problem signatures (classification), identifying unknown events (short-term analysis using outlier detection), or identifying drift over a longer period of time (long-term outlier detection).

Condition monitoring can be performed continuously or periodically. Continuous monitoring should detect a problem straight away but it is often expensive; energy hungry, which is a problem for WSNs where the network components need power; and the sensor data are very noisy, which requires careful preprocessing to ensure accurate diagnostics. Periodic monitoring is cheaper, uses less energy, and allows time for data cleaning and filtering but a problem will only be diagnosed at the next processing run. This may be acceptable for some situations that change slowly such as cracks developing in bridges but for time critical scenarios, then continuous monitoring is necessary.

In basic condition monitoring, the system is only able to distinguish between normal and abnormal conditions (no fault or fault). Beyond this, systems exhibit increasing levels of monitoring sophistication as outlined by the staircase of structural health monitoring of Lopez-Higuera *et al.*[78] described in Section I. A Level 5 system will comprise complex hardware, custom algorithms, and software to allow the diagnosis and/or the prognosis and even the solutions.

The topology of WSNs often varies over time. One very important factor in this topological variation is the mobility of the sensor nodes. Lorestani *et al.*[79] subdivided the communications network for their WSN into two: the fixed network relates to sensor nodes in fixed locations such as bridges, tunnels, and special points, whereas the movable network relates to sensor nodes attached to locomotives or rail wagons. The data for the movable network are logged with accompanying GPS coordinates. Movable (on-board) sensors can monitor the whole track length travelled by the train but only monitor the sections of the train where the sensor nodes are attached[84]. In contrast, track-mounted (fixed) sensor nodes can measure the whole train as it passes but only at specific points where the nodes are mounted on the track. This tradeoff needs to be considered when designing the node placements. The following sections analyze existing WSNs for condition monitoring on the railways and conducts analyzes from the perspective of fixed (static) and movable (dynamic) monitoring[122]. In each category, the systems are grouped according to the object(s) being monitored.

4.1. Fixed Monitoring

Many authors have investigated health monitoring of infrastructure, including rail bridges, tunnels, rail tracks and track beds, and other track infrastructure. Krüger *et al.*[70] noted that structural health monitoring is becoming ever more important as both train speeds and the axle loads exerted by trains increase. Until recently, inspection has been performed visually[70], but this only examined the structure superficially and intermittently. Additionally, the visual analysis needs to be interpreted by an expert, which can be subjective. Sensors are objective and can provide data from all of the structure (including internally) to allow the whole structure's health to be assessed and to analyze its durability and remaining life time. Wired sensors can be used for monitoring. However, wired systems are expensive, inflexible, time consuming to install, and the trains have to be stopped during installation[70]. The early fixed WSNs were simple such as attaching sensor nodes to the rails to monitor the rail temperature or low-voltage warning sensors that monitor the power supply to motors at points[15]. These systems simply generated a binary output (high or low). For the temperature sensor, a rail expert with knowledge of the ambient (stress free) rail temperatures set the alarm levels for "amber" and "red" warnings. If an alarm level was reached, then the system sent an SMS alert. Modern WSNs provide semiautomatic or automatic analysis of the sensor data to examine structural changes and to improve the durability of structures. Hence, WSN monitoring should reduce the overall maintenance costs. An overview of fixed monitoring systems is given in Table II in the Appendix.

1) Bridges

Bridges suffer structural defects exacerbated by the constant strains and vibrations of passing trains. Human inspection of bridges is difficult and much of the structure may be inaccessible. WSNs enable constant monitoring of the whole structure, including the internal structure of the bridge. The monitoring systems often comprise two units, as detailed in Section II-C. The first unit is low power. It detects the approach of a train and wakes the second unit. This second unit generates the measurements, whereas the bridge is under load (a train is crossing) to assess the bridge's health.

For an overall assessment, Krüger *et al.* [70] used a variety of MEMS to monitor structural changes in concrete including: piezo-based acoustic emission (AE) sensors for crack and fatigue detection, strain gauges to analyze static loads and stresses, accelerometers to analyze dynamic loads as trains pass, and temperature/humidity sensors to determine the ambient conditions. Grosse *et al.*[53], [54], Hay *et al.* [58], Hay [59], Ledeczi *et al.* [73], and Nair and Cai [90] used AE testing to detect fatigue cracks and other internal structural damage in bridges. The rate and intensity of AE from cracks defines the crack growth rate, which Hay *et al.*[58] categorized as inactive, active, or critically active. Ledeczi *et al.*[73] combined AE sensors to detect crack growth with strain gauges to identify changes in the forces exerted by trains both of which indicate structural problems. Bischoff *et al.*[19] and Feltrin [46] also analyzed force changes using resistance strain gauges to detect structural damage. Similarly, Sala *et al.*[106] and Sekula and Kolakowski [109] analyzed force changes using piezoelectric strain sensors mounted on the bridge trusses. Kolakowski *et al.*[67] used ultrasonic strain gauges to record elastic waves in bridge trusses due to vibrations caused by passing trains. Changes in vibrations can indicate structural problems similar to changes in forces. Chebrolu *et al.*[32] detected these changes in vibration using accelerometers. During monitoring, Kolakowski *et al.*[67], Sala *et al.* [107] and Sekula and Kolakowski [111] used an extra module: piezoelectric strain ("weigh in motion") sensors mounted on the track approaching the bridge, which weigh the train; thus, the train's weight can be factored into the structural assessment.

Kerrouche *et al.* [65] incorporated optical sensors into the fabric of bridges during bridge repairs to allow structural analysis. They used fiber Bragg grating (FBG) strain sensors with grooves cut in the bridge fabric to house the optical fibers carrying the FBG sensors. The authors concluded that the FBG monitoring needs more thorough evaluation as it was difficult to distinguish the actual strain from the noise.

2) Tunnels

Tunnels can be damp leading to rapid corrosion damage and other defects; thus, monitoring is essential. Monitoring of tunnels has many similarities to bridge monitoring. However, rail tunnels are often difficult to access and inspect. An added complication of tunnel monitoring compared with bridge monitoring is data transmission in a confined environment[88]. Data generated by sensors inside the tunnel have to be transmitted using a suitable relay mechanism (often multiple hop using relay nodes) from the various sensor nodes to the outside. Bennett *et al.*[17], Cheekiralla [33], Hoult *et al.* [62], and Stajano *et al.* [116] described WSNs deployed in the London Underground tunnels. Bennett *et al.*[17], Hoult *et al.* [62], and Stajano *et al.* [116] detected distortion using inclinometers and measured movement using linear potentiometric displacement transducers. They coupled the sensors with four relay nodes in a mesh topology to transmit the sensor data to a receiver unit in the tunnel. There is no GPRS signal near the London Underground tunnels. The authors used repeater units placed along the tunnel to transmit the data from the gateway to a computer located outside the tunnel at the top of a ventilation shaft. This transmission computer logged the data and transmitted them to the offsite compute server using a mobile-phone signal. Bennett *et al.* also analyzed tunnels in the Prague metro using the same sensor network. Here, a GPRS signal is available at the tunnel mouth; thus, they connected the gateway node to the transmission computer directly using an Ethernet cable with no requirement for repeater units, thus, reducing the chances of data loss.Cheekiralla[33] monitored deformation in the tunnel when construction work was happening nearby using off-the-shelf battery-powered pressure transducers to measure vertical displacements and the transverse deformation along the most vulnerable zone of the tunnel. Cheekiralla coupled each pressure transducer with a microcontroller, RF transmitter, analog-to-digital converter, and power conditioning hardware. The base station was located near the tunnel with a RF receiver unit connected to a laptop. The laptop had a wireless modem to periodically send data to a server.

3) Rail Tracks

Track monitoring systems also play a vital role in maintaining the safety of the railways. Monitoring bridges and tunnels (discussed in the previous paragraphs) uses sensors to identify and analyze defects (cracking and displacement) in large structures. In contrast, track monitoring involves identifying and analyzing defects in long narrow metal rails. Tracks can crack and displace like bridges but also twist and tilt (incline).Hence, track monitoring can vary from detecting settlement and twist such as that caused by nearby tunneling or excavation[112]; to measuring the forces exerted by train wheels on the tracks; to monitoring the development of cracks and structural flaws as trains pass and over the longer term.

Durham Geo Slope Indicator Company (DGSIC) [112] and Aboelela *et al.* [1] analyzed the track movements (shift) over time particularly where nearby construction work may affect the track or where track passes through vulnerable areas such as regions prone to landslides. Aboelela *et al.*[1] mounted inclinometers parallel with the tracks directly on the sleepers or in the ballast with continuous tensioned rails to monitor tilt. Likewise, DGSIC[112] mounted inclinometers parallel with the tracks to measure settlement. They also mounted inclinometers perpendicular to the track (on the ties) to monitor twist.Aboelela *et al.*[1] also plan to use inclinometers to monitor ties to detect broken ties or sabotage.

Other systems analyzed the effects of trains passing along the tracks using track-mounted strain gauges [10], [71], [94], [99], piezoelectric strain sensors [112], FBG strain sensors [41], [72], [118], [119], [127], AE sensors [7], [22], [121], [128], [135], and accelerometers [10], [18], [71]. Current sensors readings can be compared with historical sensor measurements and unusual readings or readings that match known problem signatures can be detected[7], [22], [41], [72], [118], [121], [127]. Strain gauges monitor the forces exerted by trains ("weigh in motion"), and accelerometers measure vibrations (motion of the track[9]), whereas AE sensors detect and analyze cracks. Tam *et al.* [119] installed FBG strain sensors on each side of a track to detect imbalances on the two sides of train wheels. If there is a large variation between the left and right axle loading then there is the danger of train derailment.

Monitoring the impact of passing trains is important for two reasons: it can detect track damage, but it can also detect damaged trains. Brickle *et al.*[23] detailed state-of-the-art automated systems using track-mounted sensors for wheel condition monitoring. Excessive forces, particularly where the train has a defect such as out-of-round wheels, flat wheels, or unbalanced axle loading, can crack sleepers, damage the rail head, and cause failure of rail by either growth of fractures, or fatigue and fracture of in-track welds[10]. Filograno *et al.* [47] used FBG sensors for train identification, axle counting, speed and acceleration detection, wheel imperfection monitoring, and dynamic load calculation by placing them in different positions on the track.Monitoring the forces exerted using strain sensors can identify defective axles and wheels[118] or incorrectly loaded trains [41]. Analyzing the track vibration data produced by accelerometers [16] and the AE signal transferred through the rail [7], [22], [121] from AE sensors can also detect out of round wheels as trains pass. Belotti *et al.*[16] incorporated an inductive axle-counter block into their system for assessing the train speed. The speed affects the expected vibrations; thus, the speed can be built into the wheel analysis model. Detecting wheel faults can also pinpoint trains that are at risk of causing a derailment. In[10], [71], the risk of derailment is represented by the ratio between the lateral and the vertical force (L/V ratio), which quantifies the relationship between the applied load and the deflection of the track[10].

Condition monitoring protects both the trains and the track, increases the track and train reliability and allows repair to be scheduled. If the track has been marked as defective and awaiting repair but is still usable, AE sensors can be attached to monitor the defects and ensure that they are not getting worse[135]. Track awaiting repair usually has a speed restriction enforced so, if it is monitored, this may not be required.

4) Rail Track Beds

As well as monitoring the track, it is important to analyze the rail bed [8]. The pore-pressure of water in the subsoil of rail beds increases as trains pass. Repeated pressuring can reduce the shear strength of the soil and increase the potential for bed failure. Wong *et al.*[132] used vibrating wire piezometers to measure this pore pressure. Vibrating wire piezometers are accurate and reliable. However, they have a limited reading frequency due to the vibrating wire. Wong *et al.*[132] generated data at a rate of 0.1 Hz and Hendry [60] notes that this low rate of data acquisition may miss the peak pore pressures under axle loads which occur at frequencies of 1 Hz. Konrad *et al.* [68] measured pore pressure using piezometers and wire potentiometers which generate data at 100 Hz. This higher frequency allows the analysis of pore pressure during the passage of individual trains. Other systems use a variety of sensors for monitoring. Aw[8] analyzed a railway site experiencing mud pumping problems that requires frequent (bimonthly) track maintenance. Aw used accelerometers to monitor the track vibrations, settlement probes to monitor track settlement, piezometers to measure ground water pressure, and temperature sensors. Hendry[60] used piezometers to measure the pore pressure and detect changes; extensometers to measure the relative vertical motion at various depths; and, an arm comprising jointed segments with accelerometers attached. The movement of the segments and overall shape of this arm is used to analyze the horizontal deformation of the foundation material during the passage of a train. Baum *et al.*[14] developed a WSN early warning system for rail landslides. It combined data from rain gauges to determine the rainfall level and likely saturation, tensiometers to measure the pore pressure and reflectometers to measure the soil water content. Tensiometers are able to detect changes in pore pressure rapidly to warn of slope instability, whereas reflectometers detect water content changes over time to warn of wet soil slippages.

5) Other Track Infrastructure

A variety of infrastructure is used to operate the tracks. Monitoring the condition of this infrastructure is an important part of railway condition monitoring. Fishplates are metal plates attached to the end of track sections to join and align them[51]. Fishplates can be tampered with, incorrectly maintained or the nuts and bolts may work loose. This may cause track sections to become misaligned potentially resulting in a derailment. Buggy *et al.*[27] and Ghosh *et al.* [51] proposed WSNs to monitor fishplates using FBG stain sensors [27] and piezo-resistive pressure sensors [51], respectively. If the nuts and bolts of a fishplate are loosened, then the strain and pressure on them change. The WSN can send an alert to the nearest control room.

The railways use track circuits which detect approaching trains and ensure correct train and signal operation [36], [49], [133]. Track circuits transmit a signal to a receiver via the two running rails of a track section[36]. If the signal reaches a receiver at the far end of the track section, then no trains are present. If a train is present on the section, then the signal is short circuited by the wheels, the signal does not reach the receiver and the track relay is not energized. Fukuta[49] noted that this receiver can be a wireless MEMS. Detecting current flows along the track can also pinpoint broken rails or disconnected wires. Railway points switch (move) the track between two alternative routes to ensure the train takes the correct route[63], [83], [86], [136]. The points work in conjunction with track circuits to detect and route trains. Hence, both the track circuits and the points must be fully functioning to accurately detect and route trains to ensure safe and reliable operation. Points monitoring systems often use a variety of sensors as they require multifaceted monitoring. Zhou *et al.*[136] monitored: distance, motor driving force, driving current and voltage, electrical noise, temperature and state changes, whereas Marquez *et al.*[83] measured a similar set of variables: voltage, current drawn by the electric motor, current drawn by the system as a whole, and the force in the drive bar measured with a load pin. Igarashi and Siomi[63] used magnetic lock-warp sensors to detect maladjusted switches and points irregularities by measuring displacement through temperature variations.

4.2. Movable Monitoring

WSNs are able to monitor the condition of a range of mechanics, systems, and environments using on-board sensors to measure parameters such as temperature, shocks, tilts, and humidity[138]. Thus, WSNs demonstrate a promising platform for real-time movable condition monitoring of transport systems, allowing the early detection and diagnosis of problems. Bruni *et al.*[24], Ngigi *et al.* [95] and Ward *et al.*[126] provided overviews of the opportunities available for on-board vehicle monitoring systems, which analyzed the mechanics. Measurements can all be taken during normal train service that allows in-service condition monitoring of the train chassis and mechanics and the rail track that the train is running on. The sensor nodes used for movable monitoring are often mounted on the train bogies or carriages. These locations are susceptible to high levels of vibration; thus, the sensor nodes have to be rugged or protected. Authors use strong adhesives to bond the nodes to the object being monitored and sealed protective cases to shield the sensor nodes. For example, Grudén*et al.*[55] used sensors to monitor train bogies. They used shock absorbing material around circuit boards to protect them, separated the battery from the circuit board, mounted the antenna separately from the circuit board, and used flexible cables.

One of the key issues with movable sensor monitoring is communication, due to the mobility of the sensors. Movable rail monitoring can transmit the data from the movable nodes to a static node when the static node is in range. Transferring data from the movable nodes to a static node often happens when a train stops in a station. Alternatively, the movable nodes can form a network with a base station within the network. If the nodes form a network on the train, then the base station is often mounted on the locomotive, where power is available to power the data transmission. This transmits the data over a suitable mechanism, such as satellite or GSM. Despite the issues with movable monitoring, mobile sensor nodes can achieve a higher degree of coverage and connectivity compared with static sensor nodes. As they move to a range of locations.

Object mobility (such as the movement of sensors) can be classified into uncontrolled and controlled mobility [13]. With uncontrolled mobility, a sensor node can move randomly and freely. In contrast, a node only moves along a predetermined route with controlled mobility. Vehicles on the railways can only move in limited trajectories, each train follows a planned route along tracks, which are set by points and signals and the movements are dictated by the rail timetable. Establishing WSNs for controlled mobility is easier compared with WSNs for uncontrolled mobility. If the routes are known and fixed, then the nodes and receivers can be arranged for maximum efficiency as wireless sensor nodes have limited transmission ranges. However, Chen *et al.*[34] noted that even with controlled mobility, trains can be over a mile long. If these trains are fitted with sensor nodes, then the off-train receivers have to be carefully designed and positioned to cope. An overview of movable monitoring systems is given in Table III in the Appendix.

1) Train Engines, Passenger Carriages, and Freight Wagons

Wagon or container failure could lead to damage of the rail infrastructure or the environment and could even result in loss of life; thus, it is important that the wagons, containers, and their respective contents are also monitored. Additionally, freight trains use unpowered and unwired railroad wagons, thus an on-board WSN can analyze the condition of the freight wagons[104] without power from the train. A number of authors have developed systems for analyzing train carriages, wagons, and engines. This can involve analyzing the mechanics of the vehicles or analyzing the contents where safety-critical material is being transported. Many of these systems form a mobile network in a daisy chain topology between adjacent carriages and with the engine acting as the base station[50], [81], [104], [112], [131]. The sensor data can also classify the rail wagon's response to the track and even detect potential derailments. Ward *et al.*[126] noted that on-board sensors can identify track defects including small vertical irregularities (dipped joints or voids), small lateral irregularities (kinks), larger vertical irregularities, larger lateral irregularities (misalignment), and track twist.

A number of authors analyzed the mechanics of carriages and engines using parameters such as the temperatures, strains, and forces at critical locations on the rail wagon shell and mechanics. Gao *et al.*[50], Nejikovsky, and Keller [93] and Tam *et al.*[119] used WSNs to monitor the dynamics and structural condition of both locomotives and passenger carriages, whereas Shafiullah *et al.*[112] and Wolfs *et al.* [131] monitored carriages. In [93], engine sensors analyzed the health of locomotive engines by measuring parameters, including water temperature, oil temperature, engine temperature, switch statuses, fans statuses, throttle positions, rotations per minute, and traction motor currents. Similarly, Tam *et al.*[117] used FBGs to analyze the strain and temperature of train body shells in critical locations, which include the welding joints, cross beams, and sole bars. Accelerometers can analyze different properties of the wagons depending where they are mounted[50], [93], [131]. Wolfs *et al.* [131] placed them at the edges of rail carriages; Gao *et al.* [50] placed them on the floor of the locomotive and the chassis of carriages; and Nejikovsky and Keller[93] mounted them on the body of passenger carriages of tilting trains to monitor rail wagon body motions.These train-mounted accelerometers can measure vertical and horizontal vibrations[50], [131]. They can also monitor the lateral accelerations when travelling around curves to ensure the lateral forces do not exceed government safety levels[93].Gyroscopes [131] can measure the longitudinal accelerations, as well as pitch, roll, and yaw of carriages to analyze motion.

Casola *et al.* [31], Flammini *et al.* [48] and Maly *et al.* [82] proposed to integrate data from a range of heterogeneous sensors for more holistic condition monitoring. Maly *et al.*[82] performed hot box detection, flat wheel detection, dynamic scaling, and loading gauge detection using commercially available sensors, whereas Casola *et al.*[31] analyzed temperature, humidity, and vibrations among others (the specific sensor types are not stated). The sensors are grouped into units, which operate independently. If a unit detects a faulty train, then it alerts the control center and provides detailed information about the detected irregularity.

Wagons containing freight are susceptible to tampering and intrusion. Authors have combined accelerometer data with light, shock, humidity, and temperature readings to provide security monitoring[81], [104], [112]. Mahlknecht and Masani [81] used vibration shock sensors; Mahlknecht and Masani [81] and Shafiullah *et al.* [112] monitored the humidity to detect changes; Mahlknecht and Masani [81] and Reason *et al.* [104] used light sensors to detect wagons that are open, and Mahlknecht and Masani[81] used motion detectors for analyzing movement and detecting irregularities.

2) Train Bogies

The train bogies perform a number of tasks, mainly to guide the train on both straight and curved tracks, to safeguard the train's dynamic stability, and to provide a comfortable ride for the passengers[126]. Most commercial condition monitoring systems for rail vehicles analyze the bogie system. Ngigi *et al.*[95] noted that this is because the bogie's important components undergo rapid changes during journeys and can pose safety related issues. Train axles and bogies run hot during train journeys. However, excess heat is generated as the axles and bogies wear. It is imperative to detect this excess heat to allow early detection of wear and deterioration, schedule maintenance and thus prevent vehicle failures and derailments. Hot axle bearing detectors (HABDs) have been in use for a number of years[98]. The early HABDs could detect defects but were only capable of generating an alarm once a certain threshold was reached[98]. In contrast, new systems generate a significant amount of data, both as alarms and as time-series graphs of the bearing temperatures. Different heat types have different time-series signatures[98]. Hence, it is possible to identify faults such as hot wheels caused by brake drag or extraneous heat sources such as leaking engine exhaust fumes. Previously, when only a simple threshold was used, alarms caused by these other sources of heat may have been misinterpreted as false positives, due to the lack of detailed information[98]. Rabatel *et al.* [102], [103] and Carrascal *et al.* [30] used MEMS temperature sensors; Kim *et al.* [66] used surface acoustic wave (SAW) sensors mounted on the train's bogies to detect overheated wheel bearings and Grudén*et al.*[55] used temperature sensors to analyze wheel performance. Grudén *et al.*[55] used three sensors to measure bogie temperatures and a fourth sensor to measure the air temperature to establish the ambient conditions. The ambient conditions affect the train mechanics, thus the need to be incorporated into the condition monitoring to prevent false positives and false negatives.

Bogies can also be analyzed for vibrations. Elia *et al.* [44], Mei and Li [87] and Tsunashima *et al.* [124] proposed systems for the measurement of bogie vibrations. Elia*et al.* and Tsunashima *et al.* mounted accelerometers on the bogies and Elia *et al.* also placed them on axle boxes of train carriages to measure the lateral acceleration. Mei and Li used robust inertia sensors mounted onto the bogie frame to monitor vibrations and ground speed. The papers also demonstrated that rail/wheel conicity caused by incorrect track gauge or inclination[44] and track irregularities [87], [124] can be identified from the respective bogie-based sensors. Tsunashima *et al.*[124] were able to analyze the sensor data onboard the train at source to provide real-time track analysis.

3) Train Axles

As well as generating excess heat, excess strains (such as torsional and bending stress) may be exerted on axles causing deterioration. The degree of torsional stress is affected by the train's traction control system, mechanical brake characteristics, the wheel–rail interface and the track adhesion conditions. This number of factors affecting the stress means that establishing a baseline to create a model necessitates measuring the torsional axle stress levels over a long term, ideally over the four seasons of a year[42]. Similarly, the axle bending stress levels vary due to the track quality, vehicle loading, braking and traction effects and curving speeds. Deuce and Rosinski[42] also observed strong electromagnetic interference (EMI) when measuring strains on powered axles.This required the development of bespoke instruments, including strain gauge amplifiers, telemetry systems, solid state data loggers, and signal processing units equipped with wireless data communication[42].

4) Train Wheels

A number of the systems already discussed for track and train vehicle monitoring are able to detect wheel irregularities. Train wheels wear over time and can also suffer flat wheels and out-of-round wheels. The previous section (Section IV-A fixed monitoring) discussed how WSNs attached to the track detected faulty wheels by analyzing vibrations and stresses exerted by the wheels on the tracks. Wheel problems can also be detected by sensors on board the train analyzing the wheel surfaces or analyzing forces on the train bogies and wheels. The wheel-sets and bogie each have two degrees of freedom (lateral and yaw)[75]. Wired sensors could be used for analysis but require holes in the train axle for the power leads, which weaken the axle, whereas wireless sensors are not very durable in the harsh conditions[84]; thus, this lack of durability has to be considered.

Nuffer and Bein [96] introduced a wear detection system for train wheels. It detected changes in the vibration behavior of the entire wheel caused by surface changes on the rolling contact area. The system positioned piezoelectric sensors on distinct areas of the wheel. The sensors represented the surface as an electrical signal. Changes in the surface can then be detected by changes in the electrical signal. However, it proved difficult to adequately define the correlation between the behavior of the wheel and the measured signals; thus, the system was sensitive to small changes in model parameters.Li *et al.*[75] noted that the behavior models for rail monitoring are very complex with strong interdependencies between the measured variables and between different degrees of freedom. Thus, a number of variables need to be measured. They mounted sensor nodes on the bogie frame to monitor the dynamic behavior of wheel profile (conicity) and dampers. Two accelerometers mounted on the front and rear wheel set measure the lateral accelerations of the wheel set. An accelerometer and a gyromounted on the bogie measured the lateral acceleration and yaw velocity of the bogie. Finally, a vehicle body accelerometer measured the lateral acceleration of the vehicle body.

It is also important to monitor the wheel/rail contact to detect skidding and wear.Matsumoto *et al.*[84] applied a multifaceted approach. They fitted noncontact gap sensors on nonrotating parts of a bogie to measure the bending deflection (distortion) of the wheel by the lateral contact force. They also used magnetostrictive displacement sensors to measure the deflection of the primary suspension by the vertical contact force. These two measurement combined give the wheel/rail contact forces. Matsumoto*et al.*[84] used the same derailment coefficient as [10] and [71] (discussed in the fixed monitoring section): the ratio between the lateral and the vertical force (L/V ratio).Balas and Jain[9] fused data from speed sensors and piezoelectric accelerometers attached to train wheels. Using data modeling, they were able to detect skidding if the integral of the acceleration (speed of the wheel) measured by the accelerometer differed from the speed of the carriage. They also calculated the wheel adherence as a function of the carriage weight, carriage velocity, brake cylinder pressure, and the skid amount from the skid model.

5) Train Brakes

Heavy haul transportation involves carrying heavyweight railway freight. The trains have multiple locomotives, which must coordinate their braking. Song *et al.*[115] developed a WSN for brake coordination and monitoring to reduce the maintenance and inspection requirements. They attached air pressure sensors to the locomotives and train rear, installed relay nodes on the carriage roofs, and placed accelerometers and ultrasonic sensors along the track along with railway relay nodes. When the front locomotive brakes, the sensors in it detect the change of air pressure, and the relay node on the locomotive transmits a message to the other locomotives via the relay nodes on either the train or the track using a wireless radio link. The other locomotives can then apply their brakes according to the message from the lead locomotive. The track relay nodes are used where transmission via train carriages is weak such as bends and tunnels. The data regarding the railway infrastructure from the accelerometers and ultrasonic sensors along the track can be collected and analyzed for condition monitoring.

6) Train Pantographs

In electrified railways, the overhead power lines and the pantographs on the trains require monitoring. The pantograph ensures that the train contacts the overhead power line to collect current and power the train. Ensuring the integrity of the pantograph, the pantograph/overhead line interface and the overhead power lines is important to ensure the power supply to the train, prevent breakdowns and identify sections of power line requiring maintenance. Thus, monitoring the force exerted between the contact and the wire is essential to detect and prevent problems.

Boffi *et al.* [21], Bruno *et al.* [25], Elia *et al.* [44], Waki [125], and Willsch *et al.* [129] all performed condition monitoring of the pantograph itself. Elia *et al.* also monitored the overhead power lines for irregularities and loss of tension. Optical sensors are immune to EMI making them ideal to use near power lines where EMI is high. Thus, Boffi *et al.* and Willsch *et al.* measured the strain exerted between the pantograph's current collector and the overhead wire using interferometric fiber[21] and FBG [129] sensors. Elia *et al.* and Waki used a variety of sensors to analyze different aspects of the pantograph. Both used accelerometers that can detect impact accelerations that occur when passing a contact failure and both integrated sensors in the current collector. Elia*et al.* used load cells to measure the reaction forces and displacement transducers to measure the motion of the current collector and articulated frame. Waki along with Bruno *et al.* attached ultraviolet (UV) sensors to the current collector to detect arcing. Arcing occurs when the contact between the pantograph and the overhead wire is lost and causes power transmission loss and overheating. To detect overheating further, Waki measured the pantograph's temperature using a standard temperature sensor. Waki used a video camera to record overhead line conditions and pantograph movement. Leung *et al.*[74] used passive magnetoelectric current sensors to analyze the current in the transmission cable running from the pantograph to the train's power unit. Changes in the current flow can indicate power transmission loss and need to be detected.

5. CONCLUSION

This paper has reviewed the range of WSNs used for condition monitoring in the railway industry. The emphasis is on the practical engineering solutions, principally which sensors devices are used and what they are used for; and, identification of sensor node configurations and network topologies.

Until recently, railway inspection has been visually performed [70], but this only examines objects superficially and intermittently, and the analysis needs to be interpreted by an expert, who can be subjective. Sensors are objective and can provide data from the entire object (including internally) to allow the whole object's health to be fully assessed and to analyze its durability and remaining life time. A broad range of sensors are used in railway monitoring to provide an extensive range of data and allow monitoring of different structures, vehicles and machinery. The main challenge for WSNs in railway applications is determining the best measurement technologies to use. The WSN must be reliable and accurate to enable effective condition monitoring in harsh and inaccessible environments but must also be cost effective. It must be possible to translate the sensor data from the WSN into relevant and clear information to enable decision support in the railway infrastructure maintenance lifecycle.

The paper divides railway condition monitoring into fixed monitoring for immobile infrastructures such as bridges, tunnels, tracks and associated equipment, and movable monitoring for vehicles and their mechanics. Fixed monitoring uses sensors to monitor vibrations, stresses and sound waves passed through structures (acoustics) caused by passing trains (short-term monitoring) and also changes in stresses, pressures and sound waves passed through structures over the longer term (long-term monitoring).

One of the key issues for fixed monitoring is network topology. The topology is constrained by the requirements of the monitoring and by the physical environment.Sensor nodes can be arranged in either an *ad hoc* or a preplanned configuration[134]. Determining the optimum node placement is a complex task and often requires a tradeoff. The network configuration can be optimized against a number of different constraints. A network may minimize relay nodes, may need to ensure a minimum level of service (include a certain level of redundancy), minimize energy usage to preserve battery life, or may need to ensure accessibility of the nodes[116]. The communication mechanisms, for example, Bluetooth, Wi-Fi, GSM, or satellite also need to be evaluated to ensure coverage and reliability and the routing protocol to ensure data is successfully transmitted from sensor node to base station to control center. Another issue is powering the sensors as fixed monitoring often requires placing sensor nodes in inaccessible locations, for example, in tunnels, on bridge trusses, or in rail track beds. Many sensors use batteries but replacing batteries in inaccessible locations may not be possible. Authors have considered ambient energy harvesting such as converting vibrations caused by passing trains into energy or using solar power. These energy harvesters may be accompanied with energy storage such as capacitors. Authors have also reduced sensor energy usage using event detection where a single cheap sensor detects approaching trains and wakes a larger sensor network to commence taking measurements of a bridge structure.

Movable monitoring uses sensors to monitor temperatures, vibrations, stresses, forces, currents, and sound waves in vehicles and their mechanics particularly bogies and chassis. These locations are susceptible to high levels of vibration and EMI; thus, the sensors have to be rugged or protected. One of the key issues with movable sensor monitoring is communication due to the mobility of the sensors. Movable rail monitoring either transmits the data from the movable nodes to a static node when the static node is in range, or the movable nodes form a network with a base station within the network which transmits the data over a suitable mechanism, such as satellite or GSM.

This paper focuses on the sensor technology used to generate condition monitoring data to enable practical condition monitoring systems. These data must be managed and turned into useful information to generate useful information. Condition monitoring systems must store large quantities of data to build models for analysis. The data must be validated first to ensure that they are correct and error-free (sensor faults, noise, null values, communication errors, etc.). This process may even be performed in the sensor node's microcontroller; thus, only valid data are transmitted thus minimizing the transmission load. These data can then be processed in a number of different ways to generate information. Once data are collated, they can be analyzed using robust algorithms to identify faults in near real-time. Algorithms need to be robust as WSN data is noisy, can be intermittent, may contain errors, has many interdependencies and the data volume is very high. The WSN data can also be stored and analyzed over longer time periods to identify long-term progressive faults such as a slowly developing crack.

Some systems incorporate contextual data that describe the ambient conditions, which will affect the object monitored. These contextual data can be built into models to improve the coverage and accuracy of the model and to help provide explanation of condition monitoring decisions. Elia *et al.*[44] included: train speed, position along the line (longitude and latitude from GPS system) and wind speed and direction when analyzing bogie vibrations, whereas Rabatel *et al.*[102], [103] included structural criteria (route) and environmental criteria (weather, travel characteristics, travel duration) when analyzing bogie temperatures.

6. FUTURE RESEARCH

There are a number of promising directions for further research in condition monitoring in the railways. One future direction is a move toward holistic integrated systems[31], [48], [82] which provide near real-time information and alerts. These approaches will integrate data from different sensor systems using sophisticated modeling techniques. They will also incorporate more contextual data into the modeling, including the ambient conditions, the route, the journey time, the weight of the train, and more. The models will use multistage data fusion. This fuses measurements from a variety of sources (sensors and contextual data) and over a range of time epochs to generate a consolidated state history of the object being monitored[38]. The models will also mitigate data dependencies (physical dependencies) across the physical objects being monitored. An integrated multifaceted approach should improve prediction quality. For example, train monitoring data can be combined with route data and GPS data. If multiple trains detect a vibration fault at exactly the same GPS location, then the fault is more than likely in the rail infrastructure such as uneven track. However, if only one train identifies a vibration anomaly at a particular GPS location, then the fault is more likely to be in the rolling stock [64].

Table 1 Table Detailing the Sensor Devices Used in Railway Condition Monitoring With WSNs

Sensor Type	Description and Example Usage
Accelerometers	Railway WSNs generally use **accelerometers** to measure vibrations on infrastructure such as metal tracks or bridge girders or lateral accelerations of mechanics such as wheels. They are robust, reliable, easy to calibrate and cheap. They are often used in conjunction with strain gauges [10], [71].
Displacement Transducer	A **displacement transducer** converts a linear or angular displacement into a signal suitable for recording. They can monitor movement across cracks and joints in precast concrete panels in a railway tunnel [62] to ensure the tunnel structure is stable.
Fiber-Optic Sensors, e.g., FBG (Fiber Bragg Grating) Sensors; Interferometric Fibre Sensors	**Fiber Bragg gratings** can be used as sensing elements in optical fiber sensors to measure temperature [65], strain [65] or acc leration [20]. A change of temperature or strain causes a shift in the Bragg wavelength (the wavelength of light reflected by the optical fibre). **Interferometric fibre** sensors use phase detection of the sinusoidal output power signal. The wavelength of light in optical sensors is not affected by electromagnetic fields. This immunity to electromagnetic interference is important in an electromagnetically noisy environment such as an electrified railway. FBG sensors also have long life (>20 years).
Gyroscopes (Angular Rate)	**Gyroscopes** precisely measure a vehicle's angular velocity around axes i.e., longitudinal, lateral and vertical accelerations as well as pitch, roll and yaw [131] making them ideal for analyzing carriages, chassis and bogies.
Inclinometer (Tilt Meters)	**Inclinometers** measure the slope angle by generating an artificial horizon and measuring angular tilt with respect to this horizon. Inclinometers can detect distortion of railway structures by detecting changes in incline [62].
Light (Ultra-Violet)	**Phototube** sensor is a photocathode and a series of electrodes in an evacuated glass enclosure, able to measure the duration of ultraviolet emissions (in the wavelength range 175–195 nm) due to electrical arcing [25]. They are used to analyze pantographs.
Magnetoelectric	**Magnetoelectric** sensors induce electric polarization in a material when a magnetic field is applied. They can be used to analyze magnetic fields in current-carrying electrical cables and, thus, monitor the current. They are passive so do not require an energy supply (battery) as they are powered by induction.
Piezoelectric	**Piezoelectric** sensors generate a signal when the piezoelectric material (such as tourmaline or quartz) is compressed. They are used to measure strain, pressure, vibrations and shocks. The sensors are rugged and stable over repeated use so can be used on train wheels to monitor any displacement on the wheel surface to identify wheel defects [96]. They are easier to fabricate and cheaper to run than optical sensors.
	Piezoresistive pressure sensors have a micro-machined silicon diaphragm with piezoresistive strain gauges embedded in it, fused to a silicon or glass backplate. They produce a larger signal than other sensor types providing better sensitivity. They are used to measure changes in resistance and can monitor the nuts and bolts on rail fishplates to ensure they do not work loose [51].
	Acoustic Emission (AE) sensors are high-frequency, passive piezoelectric transducers [7]. AEs are transient elastic waves generated when there is a sudden release of strain energy within a material [26], for example, the energy released by active fatigue cracks when an external force is applied. AE sensors detect these sound waves while filtering out ambient noise [59]. Thus, AE sensors are able to detect internal damage within structures where cracks are growing [53], [54], [59]. One drawback is that AE data is large and requires accurate and efficient algorithms to process and interpret it [53], [54].

	Surface Acoustic Wave (SAW) sensors convert electrical input signals into mechanical waves using a piezoelectric material. They then convert the wave back into an electrical signal for analysis. Mechanical waves are affected by physical phenomena such as temperature so SAW sensors are used to monitor temperature changes in the axle box bearing of railroad vehicles [66]. The sensors are very sensitive and stable but the waves are distorted by liquids so SAW sensors need to be dry.
Pressure Transducer, Piezometer, Time Domain Reflectometer, Settlement Probe, Tensiometer,	A **pressure transducer** produces a signal to reflect the pressure imposed on the sensor. Pressure transducers can determine vertical displacement [33] from a hydraulic reference line. Any displacement will be reflected by changes in pressure. **Piezometers** have been used in the railways to measure the positive water pressure in the ground water of rail beds [8]. They are hydraulic pressure transducers where external water exerts pressure on internal fluid which in turn exerts pressure on an internal pressure transducer. They are robust and reliable but need to be calibrated to account for atmospheric pressure. **Time Domain Reflectometer (TDR)** converts the travel time of a high frequency, electromagnet pulse into volumetric water content. They are simple to use and easy to calibrate. **Settlement probes** measure the long-term settlement of rail tracks on the rail bed [8]. They are hydraulic pressure transducers with an embedded probe able to measure large settlement. As the probe changes elevation then the pressure changes which is measured by the pressure transducer. They are sensitive to the ambient (atmospheric) pressure and need to be calibrated carefully. **Tensiometer** measure negative water pressure (water tension) in the soil [8] to quantify the soil moisture status. The water exerts a negative pressure on internal fluid and thus on an internal pressure transducer in the sensor. They require regular maintenance. **Vibrating wire piezometers** measures changes in ground water pressure by measuring the strain exerted by the water acting on a diaphragm and monitoring the vibrations in a wire touching the diaphragm. They are durable, stable and reliable but the vibrating wires limit the frequency of readings to several seconds per reading [60] (e.g., 0.1Hz [132]) so they cannot analyze changes in pore pressure during the passage of individual trains. They are suited to long-term monitoring.
Strain Gauges	**Strain gauges** measure local stresses and generate a signal to reflect the strain applied to the sensor. They are cheap, simple and accurate. A resistance strain gauge contains metallic foil which deforms as strain is applied causing its electrical resistance to change. A vibration wire strain gauge contains a steel wire held in tension. The sensor generates a signal reflecting the strain on the sensor as a function of the tension in the wire. Strain gauges often measure the vertical and lateral force applied to the rail between two sleepers (or ties) [10], [71], [99] as trains pass. When a train is present the strain increases on the rail, which is detected by the sensors.
Temperature Sensors (Thermocouples)	In railway applications, **temperature sensors** are used to monitor the temperature of: the atmosphere (air temperature), the rail bed [8] or train chassis and mechanics [55], [102], [103], [104], [105].
Ultrasonic (e.g., Strain Gauges)	**Ultrasonic strain gauges** record elastic waves in a structure due to vibrations caused, for example, by a passing train [67]. They are easier to install and are more sensitive than standard strain gauges and can be relocated unlike piezoelectric strain gauges [67].

Table 2 Table Listing the Fixed Objects Monitored, the Measurements Used for Each Object, the Sensor Devices Used to Monitor it and a List of Citations to Systems Monitoring That Object

Object monitored	Measurement	Sensor	Citations
Bridge	Crack/Fatigue Detection	Acoustic Emission	[53] [54] [58] [59] [70] [73]
	Stresses	Strain Gauge	[19] [46] [70] [123]
		Piezoelectric Strain Gauge	[107][111]
		Fiber Bragg Strain Gauge	[65]
		Ultrasonic Strain	[67]
	Weight (of Train)	Strain Gauge	[106]
	Vibrations (Dynamic Load)	Accelerometer	[32] [70] [107]
Tunnel	Structure Distortion	Inclinometers	[17] [62] [116]
	Structure Movement	Displacement Transducers	[17] [62] [116]
	Vertical Displacement	Pressure Transducer	[33]
	Transverse Deformation	Pressure Transducer	[33]
Track	Crack/Fatigue Detection	Acoustic Emission	[7] [22] [121] [128] [135]
	Out of Round Wheel	Acoustic Emission	[7] [22] [121]
		Accelerometer	[16]
	Stresses	Strain Gauge	[9] [67] [71] [94] [99] [111]
		Piezoelectric Strain Gauge	[67][109]
		Fiber Bragg Strain Gauge	[41] [47] [72] [118][119] [127]
	Vibrations (Dynamic Load)	Accelerometer	[10][18] [17] [71]
		Fiber Bragg	[47]
	Settlement and Twist	Inclinometer	[112]
	Incline	Inclinometer	[1]
Rail bed	Dynamic Acceleration (Track)	Accelerometers	[8]
	Pore Pressure (Ground Water)	Piezometers	[8] [60] [68] [132]
	Pore Pressure (Ground Water)	Tensiometers	[14]
	Pore Pressure	Wire Potentiometers	[68]
	Long Term Settlement (Track)	Settlement Probes	[8]
	Temperature	Temperature Sensors (Thermistor)	[8]
	Vertical Motion	Extensometer	[60]
	Water Content	Reflectometer	[14]
	Movement / shape (of rail bed)	Accelerometers mounted in arm	[60]
Track infrastructure	Pressure	Piezoresistive Pressure Sensors	[51]
	Strain	Fiber Bragg Strain Gauge	[27]
	Displacement	Magnetic	[63]

As sensors become cheaper and easier to fabricate, then their availability and use will increase. Bukova and Svecova [28] proposed a WSN comprising many tiny wireless MEMS (smart dust), which are able to collect, process, and transmit data from light to vibrations. Vast numbers of these sensors can be placed inside a wagon, for example, where they can measure temperature, humidity, or vibrations. Bukova and Svecova posited that recent developments in sensor fabrication techniques will produce MEMS that are the size of a grain of sand yet contain computing circuits, bidirectional wireless communications technology and a power supply.

The number of sensors will increase in the future, and the number and distribution of monitored objects will expand. This will require further development of energy efficient routing and transmission mechanisms to ensure reliable and consistent communication. Condition monitoring will also tend to parallel and distributed processing across multiple nodes to speed execution. This could be processing the data in parallel at the same geographical location using a single machine with multiple processing cores or at the same geographical location using multiple compute nodes or even distributed processing, where the data are processed at multiple geographical locations, and the results are assimilated at a central location.

As WSNs expand and become more pervasive, security against intrusion, eavesdropping, data tampering and unauthorized control will be ever more important[116]. WSNs pose unique problems for security. WSNs broadcast data between nodes making them vulnerable. The sensors are energy constrained with limited processing capability; thus, any security software needs to be carefully designed.Network administrators will have to identify which WSNs need to be secured, how they need to be secure, and what level of security is required[117]. The type of security is important. Does the WSN need to be secured for access? How often and to what level should the WSN be monitored for intrusion detection? Does the data need protecting from data tampering? The level of security will be dictated by the value of the asset, the type of asset and the consequences of a failure. WSNs will become more interlinked; thus, it is imperative that WSN security cannot be compromised through access from other networks.

In the future, ever more data will be collected from railway infrastructure and vehicles.This inevitably leads to a requirement for the railway industry to develop standards for data collection[126] and processing. West *et al.* [110] proposed developing Web Ontology Language standardized ontologies to encapsulate railway data and ensure consistency for transmission and processing. Different systems and different companies can then share data using a consistent interface.

Table 3 Table Listing the Movable Objects Monitored, the Measurements Used, the Sensor Devices Used and a List of Citations to Systems Monitoring That Object

Object monitored	Measurement	Sensor	Citations
Train Shell	Strain	FB Strain	[119]
	Temperature	FB Temperature	[119]
		Temperature Sensors	[29] [102] [103]
	Vibrations	Accelerometers	[50] [113] [131]
	Lateral Acceleration	Accelerometer	[93]
	Longitudinal Accelerations, Pitch, Roll And Yaw	Gyroscopes	[113] [131]
Wagon	Temperature	Temperature Sensors	[81] [104] [105] [112]
	Humidity	Humidity Sensors	[81] [112]
	Vibrations /Acceleration	Accelerometer	[81] [104] [105] [112]
	Vibrations	Vibration	[81]
	Light	Light Sensor	[81] [104] [105]
	Movement	Motion Detector	[81]
Bogies	Temperature	Temperature Sensors	[55] [98]
		Surface Acoustic Wave	[66]
	Vibration	Accelerometers	[44] [87] [124]
		Inertia Sensors	[87]
Axles	Stress	Strain Gauges	[42]
	Temperature	Thermocouples	[42]
		SAW temperature	[109]
	Vibration	Accelerometers	[91]
Wheels	Vibration Across Surface	Piezoelectric	[96]
	Lateral Acceleration	Accelerometer	[44] [75]
	Wheel Acceleration	Accelerometer	[9]
	Lateral Contact Force	Gap Sensors	[84]
	Vertical Contact Force	Magnetostrictive Displacement Sensors	[84]
	Yaw	Gyro	[75]
Brakes	Wind Pressure	Wind Pressure Sensors	[115]
Pantograph	Strain	Fiber Bragg Strain	[129]
	Temperature	Fiber Bragg Temperature	[129]
		Temperature Sensor	[125]
	Impact Acceleration	Accelerometer	[44] [97] [125]
		FBG Accelerometer	[20]
	Motion	Displacement Transducer	[44]
	Contact (Reaction) Force	Load Cell	[44]
	Current flow	Magnetoelectric current	[74]
	Arcing	UV sensor or phototube	[25] [125]

As more and varied condition monitoring systems are developed, then there is a further requirement of standardization of presentation of decisions and information across these systems for consistency and to allow integration of multiple systems. Equally, many condition monitoring algorithms are "black boxes" providing little or no explanation of decisions[110]. In the staircase of structural health monitoring of Lopez-Higuera *et al.*[78], a Level 4 system can estimate the consequences of the damage and remaining service life. However, only a Level 5 system will allow the diagnosis and/or the prognosis and recommendation of a solution to the problem. Future systems will need to provide clearer and more detailed explanation of their recommendations to increase usability.

In the future, condition monitoring will be able to exploit cheaper, more robust and more pervasive hardware. New data processing techniques will generate more accurate, robust, and reliable models from existing and new sensor data. Data and processing will be standardized. Systems will be secure and provide clear and detailed decisions and recommendations.

ACKNOWLEDGMENT

The authors would like to thank the reviewers for their helpful suggestions, which improved the quality of this review.

REFERENCES

1. E. Aboelela, W. Edberg, C. Papakonstantinou and V. Vokkarane "Wireless sensor network based model for secure railway operations" *Proc. 25th IEEE Int. Perform., Comput. Commun. Conf., pp. 1-6, 2006*

2. M. Aguado "WiMax on rails: A broadband communication architecture for CBTC systems" *IEEE Veh. Technol. Mag., vol. 3, no. 3, pp. 47-56, 2008*

3. B. Ai "Challenges toward wireless communications for high-speed railway" *IEEE Trans Intell. Transp. Syst., vol. 15, no. 5, pp. 2143-2158, 2014*

4. K. Akkaya and M. Younis "A survey on routing protocols for wireless sensor networks" *Ad Hoc Netw., vol. 3, no. 3, pp. 325-349, 2005*

5. Akyildiz, W. Su, Y. Sankarasubramaniam and E. Cayirci "Wireless sensor networks: A survey" *Comput. Netw., vol. 38, no. 4, pp. 393-422, 2002*

6. Akyildiz, W. Su, Y. Sankarasubramaniam and E. Cayirci "A survey on sensor networks" *IEEE Commun. Mag., vol. 40, no. 8, pp. 102-114, 2002*

7. Anastasopoulos, K. Bollas, D. Papasalouros and D. Kourousis "Acoustic emission on-line inspection of rail wheels" *Proc. 29th Eur. Conf. Acoust. Emission Testing, pp. 1-8, 2010*

8. E. Aw Low cost monitoring system to diagnose problematic rail bed: case study of mud pumping site *2007, Massachusetts Inst. Technol.*

9. V. Balas and L. Jain "World knowledge for sensors and estimators by models and internal models" *J. Intell. Fuzzy Syst., vol. 21, no. 1, pp. 79-88, 2010*

10. D. Barke and W. Chiu "Structural health monitoring in the railway industry: A review" *Struct. Health Monitoring, vol. 4, no. 1, pp. 81-93, 2005*

11. P. Baronti "Wireless sensor networks: A survey on the state of the art and the 802.15. 4 and ZigBee standards" *Comput. Commun., vol. 30, no. 7, pp. 1655-1695, 2007*

12. G. Barrenetxea, F. Ingelrest, G. Schaefer and M. Vetterli "The hitchhiker's guide to successful wireless sensor network deployments" *Proc. 6th ACM Conf. Embedded Netw. Sens. Syst., pp. 43-56, 2008*

13. S. Basagni, A. Carosi and C. Petrioli "Controlled vs. uncontrolled mobility in wireless sensor networks: Some performance insights" *Proc. IEEE 66th Veh. Technol. Conf., pp. 269-273, 2007*

14. R. Baum "Early warning of landslides for rail traffic between Seattle and Everett" *Proc. Landslide Risk Manage., pp. 731-740, 2005*

15. C. Bell "Event monitoring comes of age: ?predict and prevent? as well as ?find and fix?" *Proc. IET Int. Conf. Railway Condition Monitor., pp. 44-47, 2006*

16. V. Belotti, F. Crenna, R. Michelini and G. Rossi "Wheel-flat diagnostic tool via wavelet transform" *Mech. Syst. Signal Process., vol. 20, no. 8, pp. 1953-1966, 2006*

17. P. Bennett "Wireless sensor networks for underground railway applications: Case studies in Prague and London"*Smart Struct. Syst., vol. 6, no. 5/6, pp. 619-639, 2010*

18. E. Berlin and K. van Laerhoven "Sensor networks for railway monitoring: Detecting trains from their distributed vibration footprints" *IEEE Int. Conf. Distrib. Comput. Sens. Syst., pp. 80-87, 2013*

19. R. Bischoff, J. Meyer, O. Enochsson, G. Feltrin and L. Elfgren "Event-based strain monitoring on a railway bridge with a wireless sensor network" *Proc. 4th Int. Conf. Struct. Health Monitor. Intell. Infrastructure, pp. 1-8, 2009*

20. M. Bocciolone, G. Bucca, A. Cigada, A. Collina and L. Comolli "An application of FBG accelerometers for monitoring pantographs of underground trains" *Proc. SPIE 4th Eur. Workshop Opt. Fibre Sens., pp. 765341-1-765341-4, 2010*

21. P. Boffi "Optical fiber sensors to measure collector performance in the pantograph-catenary interaction" *IEEE Sens. J., vol. 9, no. 6, pp. 635-640, 2009*

22. K. Bollas, P. Dimitrios, K. Dimitrios and A. Athanasios "Acoustic emission inspection of rail wheels" *J. Acoust. Emission, vol. 28, pp. 215-228, 2010*

23. B. Brickle Identification of Existing and New Technologies for Wheelset Condition Monitoring *2008*

24. S. Bruni, R. Goodall, T. Mei and H. Tsunashima "Control and monitoring for railway vehicle dynamics" *Veh. Syst. Dyn., vol. 45, no. 7/8, pp. 743-779, 2007*

25. O. Bruno, A. Landi, M. Papi and L. Sani "Phototube sensor for monitoring the quality of current collection on overhead electrified railways" *Proc. Inst. Mech. Eng. Part F: J. Rail Rapid Transit, vol. 215, no. 3, pp. 231-241, 2001*

26. K. Bruzelius "An initial investigation on the potential applicability of Acoustic Emission to rail track fault detection" *NDT E Int., vol. 37, no. 7, pp. 507-516, 2004*

27. S. Buggy "Intelligent infrastructure for rail and tramways using optical fibre sensors" *IET Railway Young Professionals' Best Paper Competition, 2011*

28. B. Bukova and Z. ?vecov? "Smart dust in transport services" *Railway Transp. Logist., vol. 3, no. 2, pp. 39-43, 2007*

29. Cardenas"Attacks against process control systems: Risk assessment, detection, response" *Proc. 6th ACM Symp. Inf., Comput. Commun. Security, pp. 355-366, 2011*

30. Carrascal, A. Diez and A. Azpeitia "Unsupervised methods for anomalies detection through intelligent monitoring systems" *Proc. 4th Int. Conf. Hybrid Artif. Intell. Syst., pp. 137-144, 2009*

31. V. Casola "Freight train monitoring: A case-study for the pSHIELD project" *Proc. 6th Int. Conf. Innov. Mobile Internet Serv. Ubiquitous Comput., pp. 597-602, 2012*

32. K. Chebrolu, B. Raman, N. Mishra, P. Valiveti and R. Kumar "Brimon: A sensor network system for railway bridge monitoring" *Proc. 6th Int. Conf. Mobile Syst., Appl. Serv., pp. 2-14, 2008*

33. S. Cheekiralla "Wireless sensor network-based tunnel monitoring "*Proc. 1st Workshop Real-World WSN Poster Session, 2005*

34. H. Chen "Towards scalable deployment of sensors and actuators for industrial applications" *IEEE Int. Conf. Sens. Netw., Ubiquitous Trustworthy Comput., pp. 285-290, 2008*

35. J. Chen "RAISE: RAIlway infrastructure health monitoring using wireless sensor networks" *Sens. Syst. Softw., vol. 122, pp. 143-157, 2013*

36. J. Chen and C. Roberts "Effective condition monitoring of line side assets" *IET Int. Conf. Railway Condition Monitoring, pp. 78-83, 2006*

37. C.-Y. Chong and S. Kumar "Sensor networks: Evolution, opportunities, challenges" *Proc. IEEE, vol. 91, no. 8, pp. 1247-1256, 2003*

38. S. Coraluppi and C. Carthel "Multi-stage data fusion and the MSTWG TNO datasets" *Proc. 12th Int. Conf. Inf. Fusion, pp. 1552-1559, 2009*

39. R. da Costa Marques Pimentel "Hybrid fiber-optic/electrical measurement system for characterization of railway traffic and its effects on a short span bridge" *IEEE Sens. J., vol. 8, no. 7, pp. 1243-1249, 2008*

40. N. Dadashi. "A framework of data processing for decision making in railway intelligent infrastructure" *Proc. IEEE 1st Int. Multi-Disciplinary Conf. Cognitive Methods Situation Awareness Decision Support, pp. 276-283, 2011*

41. G. den Buurman and A. Zoeteman "Quo Vadis: Review of the contribution of a weigh-in-motion system to optimising railway performance" *Proc. 7th World Congr. Railway Res., pp. 1-10, 2006, [cdrom] Available:*

42. R. Deuce and J. Rosinski "Ultra miniature data loggers for testing and in-service monitoring of railway axles and other rail vehicle components" *Proc. IET Int. Conf. Railway Condition Monitor., pp. 57-60, 2006*

43. K. Dierkx The Smarter Railroad. Transport R&D for Innovation Project Report *2009*

44. M. Elia "Condition monitoring of the railway line and overhead equipment through onboard train measurement-an Italian experience" *Proc. IET Int. Conf. Railway Condition Monitor., pp. 102-107, 2006*

45. D. Estrin "Connecting the physical world with pervasive networks" *IEEE Pervasive Comput., vol. 1, no. 1, pp. 59-69, 2002*

46. G. Feltrin "Wireless sensor networks: A monitoring tool for improving remaining lifetime estimation" *Civil Struct. Health Monitoring Workshop, pp. 1-8, 2012*

47. M. Filograno "Real-time monitoring of railway traffic using fiber Bragg grating sensors" *IEEE Sens. J., vol. 12, no. 1, pp. 85-92, 2012*

48. F. Flammini "Towards wireless sensor networks for railway infrastructure monitoring" *Proc. Electr. Syst. Aircraft, Railway Ship Propulsion, pp. 1-6, 2010*

49. Y. Fukuta "Possibility of sensor network applying for railway signal system"*Proc. 5th Int. Conf. Netw. Sens. Syst., pp. 170-172, 2008*

50. Gao"Design of train ride quality testing system based on wireless sensor network" *Proc. Int. Conf. Electron. Mech. Eng. Inf. Technol., pp. 2636-2639, 2011*

51. K. Ghosh "Development of fishplate tampering detection system for railway security based on wireless sensor network" *Int. J. Comput. Appl., vol. 2, pp. 29-34, 2011*

52. R. Goodall and C. Roberts "Concepts and techniques for railway condition monitoring" *IET Int. Conf. Railway Condition Monitoring, pp. 90-95, 2006*

53. Grosse "Wireless acoustic emission sensor networks for structural health monitoring in civil engineering" *Proc. Eur. Conf. Non-Destructive Testing, pp. 1-8, 2006*

54. Grosse, S. Glaser and M. Kruger "Initial development of wireless acoustic emission sensor Motes for civil infrastructure state monitoring"*Smart Struct. Syst., vol. 6, no. 3, pp. 197-209, 2010*

55. M. Grud?n, A. Westman, J. Platbardis, P. Hallbjorner and A. Rydberg "Reliability experiments for wireless sensor networks in train environment"*Proc. Eur. Wireless Technol. Conf., pp. 37-40, 2009*

56. Hall and J. Llinas "An introduction to multisensor data fusion" *Proc. IEEE, vol. 85, no. 1, pp. 6-23, 1997*

57. J. Han and M. Kamber Morgan Kaufmann Series in Data Management Systems *2006, Elsevier*

58. T. Hay "Transforming bridge monitoring from time-based to predictive maintenance using acoustic emission MEMS sensors and artificial intelligence" *Proc. 7th World Congr. Railway Res., 2006*

59. T. Hay Wireless Remote Structural Integrity Monitoring for Railway Bridges *2007*

60. M. Hendry The geomechanical behaviour of peat foundations below rail-track structures *2012, Univ. of Saskatchewan*

61. V. Hodge Outlier and Anomaly Detection: A Survey of Outlier and Anomaly Detection Methods *2011, Academic*

62. N. Hoult "Wireless sensor networks: Creating ?smart infrastructure?" *Proc. ICE-Civil Eng., vol. 162, no. 3, pp. 136-143, 2009*

63. Y. Igarashi and S. Siomi "Development of monitoring system for electric switch machine" *Quarterly Rep. Railway Tech. Res. Inst., vol. 47, no. 2, pp. 78-82, 2006*

64. K. Ishida, H. Kitabayashi, M. Nagasu and K. Bekki "New train control and information services utilizing broadband networks" *Hitachi Rev., vol. 53, no. 1, pp. 21-24, 2004*

65. Kerrouche "Field tests of fibre Bragg grating sensors incorporated into CFRP for railway bridge strengthening condition monitoring" *Sens. Actuators A, Phys., vol. 148, no. 1, pp. 68-74, 2008*

66. J. Kim, K. S Lee and J. Oh "A study on the wireless onboard monitoring system for railroad vehicle axle bearings using the SAW sensor" *Sens. Syst. Softw., vol. 57, pp. 52-58, 2011*

67. P. Kolakowski"Structural health monitoring of a railway truss bridge using vibration-based and ultrasonic methods"*Smart Mater. Struct., vol. 20, no. 3, p. 035016, 2011*

68. J. Konrad, S. Grenier and P. Garnier "Influence of repeated heavy axle loading on peat bearing capacity" *Proc. 60th Can. Geotech. Conf., pp. 1551-1558, 2007*

69. M. Kroener "Energy harvesting technologies: Energy sources, generators and management for wireless autonomous applications" *Proc. 9th Int. Multi-Conf. Syst., Signals Devices, pp. 1-4, 2012*

70. M. Kr?ger Sustainable Bridges. Technical Report on Wireless Sensor Networks using MEMS for Acoustic Emission Analysis including other Monitoring Tasks *2007, European Union*

71. R. Lagneb?ck Evaluation of wayside condition monitoring technologies for condition-based maintenance of railway vehicles *2007, Lule? Univ. of Technol.*

72. C. Lai "Development of a fiber-optic sensing system for train vibration and train weight measurements in Hong Kong" *J. Sensors, 2012*

73. Ledeczi "Wireless acoustic emission sensor network for structural monitoring" *IEEE Sens. J., vol. 9, no. 11, pp. 1370-1377, 2009*

74. Leung, S. Y. Zhang, S. W. Or, S. L. Ho and K. Y. Lee "Magnetoelectric smart current sensors for wireless condition monitoring of train traction systems" *Proc. 1st Int. Workshop High-Speed Intercity Railways, pp. 319-327, 2012*

75. P. Li "Estimation of railway vehicle suspension parameters for condition monitoring" *Control Eng. Practice, vol. 15, no. 1, pp. 43-55, 2007*

76. R. Liu, Y. Wu, I. Wassell and K. Soga "Frequency diversity measurements at 2.4 GHz for wireless sensor networks deployed in tunnels" *Proc. IEEE 20th Int. Symp. Pers., Indoor Mobile Radio Commun., pp. 2990-2994, 2009*

77. R. Liu, I. Wassell and K. Soga "Relay node placement for wireless sensor networks deployed in tunnels" *Proc. IEEE 6th Int. Conf. Wireless Mobile Comput., Netw. Commun., pp. 144-150, 2010*

78. J. Lopez-Higuera, L. Rodriguez Cobo, A. Quintela Incera and A. Cobo

79. "Fiber optic sensors in structural health monitoring"*J. Lightw. Technol., vol. 29, no. 4, pp. 587-608, 2011*

80. Lorestani, S. A. Mousavi and R. Ebadaty"Monitoring rail traffic using Wireless Sensor Network (WSN)"*Int. J. Comput. Sci. Eng. Technol., vol. 2, no. 6, pp. 1280-1282, 2012*

81. J. Lynch and K. Loh "A summary review of wireless sensors and sensor networks for structural health monitoring"*Shock Vibration Dig., vol. 38, no. 2, pp. 91-130, 2006*

82. S. Mahlknecht and S. Madani"On architecture of low power wireless sensor networks for container tracking and monitoring applications "*Proc. 5th IEEE Int. Conf. Ind. Inf., pp. 353-358, 2007*

83. T. Maly, M. Rumpler, H. Schweinzer and A. Schoebel"New development of an overall train inspection system for increased operational safety"*Proc. IEEE Intell. Transp. Syst., pp. 188-193, 2005*

84. Marquez, P. Weston and C. Roberts"Failure analysis and diagnostics for railway trackside equipment"*Eng. Failure Anal., vol. 14, no. 8, pp. 1411-1426, 2007*

85. Matsumoto"Continuous observation of wheel/rail contact forces in curved track and theoretical considerations"*Veh. Syst. Dyn., vol. 50, pp. 349-364, 2012*

86. S. McConnell and D. Skillicorn"A distributed approach for prediction in sensor networks"*Proc. 1st Int. Workshop Data Mining Sens. Netw.a, pp. 28-37, 2005*

87. M. McHutchon, W. J. Staszewski and F. Schmid"Signal processing for remote condition monitoring of railway points"*Strain, vol. 41, no. 2, pp. 71-85, 2005*

88. T. Mei and H. Li"Measurement of vehicle ground speed using bogie-based inertial sensors"*Proc. Inst. Mech. Eng. Part F, J. Rail Rapid Transit, vol. 222, no. 2, pp. 107-116, 2008*

89. L. Mottola"Not all wireless sensor networks are created equal: A comparative study on tunnels"*ACM Trans. Sens. Netw., vol. 7, no. 2, 2010*

90. C. Nagode, M. Ahmadian and S. Taheri"Motion-based energy harvesting devices for railroad applications"*Proc. Am. Soc. Mech. Eng. Joint Rail Conf., pp. 267-271, 2010*

91. Nair and C. Cai"Acoustic emission monitoring of bridges: Review and case studies"*Eng. Struct., vol. 32, no. 6, pp. 1704-1714, 2010*

92. W. Nan, M. Qingfeng, Z. Bin, L. Tong and M. Qinghai"Research on linear wireless sensor networks used for online monitoring of rolling bearing in freight train"*J. Phys., Conf. Ser., vol. 305, no. 1, p. 012024, 2011*

93. C. Nelson, S. Platt, D. Albrecht, V. Kamarajugadda and M. Fateh"Power harvesting for railroad track health monitoring using piezoelectric and inductive devices"*Proc. SPIE Active Passive Smart Struct. Integrated Syst., vol. 6928, no. 1, pp. 1-9, 2008*

94. B. Nejikovsky and E. Keller"Wireless communications based system to monitor performance of rail vehicles"*Proc. IEEE/ASME Joint Railroad Conf., pp. 111-124, 2000*

95. N. Nenov, E. N. Dirmitrov, G. S. Mihov, T. G. Ruzhekov and P. M. Piskulev"A study on sensors for measuring load of railway vehicle wheels in motion"*Proc. 31st Int. Spring Seminar Electron. Technol., pp. 550-555, 2008*

96. R. Ngigi"Modern techniques for condition monitoring of railway vehicle dynamics"*J. Phys., Conf. Ser., vol. 364, no. 1, 2012*

97. J. Nuffer and T. Bein"Application of piezoelectric materials in transportation industry"*Proc. Global Symp. Innovative Solutions Advancement Transp. Ind., pp. 1-11, 2006*

98. C. O'Donnell"Pantograph damage and wear monitoring system"*Proc. IET Int. Conf. Railway Condition Monitoring, pp. 178-181, 2006*

99. B. Ollier"Intelligent infrastructure?The business challenge: Railway Condition Monitoring"*Proc. IET Int. Conf. Railway Condition Monitoring, pp. 1-6, 2006*

100. M. PaloCondition monitoring of railway vehicles: a study on wheel condition for heavy haul rolling stock*2012, Lule? Univ. Technology*

101. Pascale, N. Varanese, G. Maier and U. Spagnolini"A wireless sensor network architecture for railway signalling"*Proc. 9th Italian Netw. Workshop, pp. 1-4, 2012*

102. PourghodratEnergy harvesting systems design for railroad safety*2011, Univ. Nebraska-Lincoln*

103. J. Rabatel, S. Bringay and P. Poncelet

104. "SO_MAD: SensOr mining for anomaly detection in railway data"*Adv. Data Mining Appl. Theor. Aspects, LNCS, vol. 5633, pp. 191-205, 2009*

105. J. Rabatel, S. Bringay and P. Poncelet"Anomaly detection in monitoring sensor data for preventive maintenance"*Exp. Syst. Appl., vol. 38, no. 6, pp. 7003-7015, 2011*

106. J. Reason, H. Chen, R. Crepaldi and S. DuriMobile Computing, Applications, Services*vol. 35, pp. 72-91, 2010, Springer-Verlag*

107. J. Reason and R. Crepaldi"Ambient intelligence for freight railroads"*IBM J. Res. Develop., vol. 53, no. 3, pp. 1-14, 2009*

108. M. Reyer, S. Hurlebaus, J. Mander and O. E. Ozbulut"Design of a wireless sensor network for structural health monitoring of bridges"*Proc. 5th Int. Conf. Sens. Technol., pp. 515-520, 2011*

109. D. Sala, J. Motylewski and P. Koaakowsk"Wireless transmission system for a railway bridge subject to structural health monitoring"*Diagnostyka, vol. 50, no. 2, pp. 69-72, 2009*

110. Salaberria, R. Carballedo, U. Gutierre and A. Perallos"Wireless communications architecture for ?train-to-earth? communication in

the railway industry"*Proc. 10th Int. Work-Conf. Artif. Neural Netw., pp. 626-633, 2009*

111. Scholl"SAW-based radio sensor systems for short-range applications"*IEEE Microw., vol. 4, no. 4, pp. 68-76, 2003*

112. J. Schwarzenbach, L. Wilkinson, L. West and M. PillingMapping the remote condition monitoring architecture?Core report*2010*

113. K. Sekula and P. Kolakowski"Piezo-based weigh-in-motion system for the railway transport"*Struct. Control Health Monitoring, vol. 19, no. 2, pp. 199-215, 2012*

114. Shafiullah, A. Gyasi-Agyei and P. Wolfs"Survey of wireless communications applications in the railway industry"*Proc. 2nd Int. Conf. Wireless Broadband Ultra Wideband Commun., p. 65, 2007*

115. G. Shafiullah, A. B. M. S. Ali, A. Thompson and P. J. Wolfs"Predicting vertical acceleration of railway wagons using regression algorithms"*IEEE Trans. Intell. Transp. Syst., vol. 11, no. 2, pp. 290-299, 2010*

116. G. Shafiullah, S. A. Azad and A. B. M. S. Ali"Energy-efficient wireless MAC protocols for railway monitoring applications"*IEEE Trans. Intell. Transp. Syst., vol. 14, no. 2, pp. 649-659, 2013*

117. Song, J. Zhong and X. Li"WHHT: A wireless sensor network for heavy haul transportation"*Proc. 6th Int. Conf. Wireless Commun. Netw. Mobile Comput., pp. 1-4, 2010*

118. F. Stajano"Smart bridges, smart tunnels: Transforming wireless sensor networks from research prototypes into robust engineering infrastructure"*Ad Hoc Netw., vol. 8, no. 8, pp. 872-888, 2010*

119. G. Stoneburner, A. Goguen and A. Feringa Risk management guide for information technology systems*vol. 800, 2002*

120. Gotcha?wheel flat detection and axle load measurement system*2005*

121. H. Tam"Utilization of fiber optic Bragg Grating sensing systems for health monitoring in railway applications"*Struct. Health Monitoring, Quantification, Validation, Implementation, vol. 1?2, pp. 1824-1831, 2007*

122. H. Tan, P. W. Q. Lee, W. K. G. Seah and A. Zhi"Impact of power control in Wireless Sensor Networks Powered By Ambient Energy Harvesting (WSN-HEAP) for railroad health monitoring"*Proc. Int. Conf. Adv. Inf. Netw. Appl. Workshops, pp. 804-809, 2009*

123. N. Thakkar, J. A. Steel and R. L. Reuben "Rail-wheel interaction monitoring using Acoustic Emission: A laboratory study of normal rolling signals with natural rail defects" *Mech. Syst. Signal Process., vol. 24, no. 1, pp. 256-266, 2010*

124. S. Tilak"A taxonomy of wireless micro-sensor network models"*ACM SIGMOBILE Mobile Comput. Commun. Rev., vol. 6, no. 2, pp. 28-36, 2002*

125. Townsend, S. Arms and J. Wilson Sensor Technology Handbook *2005, Elsevier*

126. H. Tsunashima, T. Kojima, Y. Marumo, H. Matsumoto and T. Mizuma"Condition monitoring of railway track using in-service vehicle"*Proc. 4th IET Int. Conf. Railway Condition Monitoring, pp. 1-6, 2008*

127. H. Waki"Monitoring of contact line equipment by trains in operation"*JR East Tech. Rev., East Japan Railway Culture Found., vol. 17, pp. 11-14, 2010*

128. Ward"Condition monitoring opportunities using vehicle-based sensors"*Proc. Inst. Mech. Eng. Part F, J. of Rail Rapid Transit, vol. 225, no. 2, pp. 202-218, 2011*

129. Wei"Real-time train wheel condition monitoring by fiber Bragg grating sensors"*Int. J. Distrib. Sens. Netw., vol. 2012, pp. 1-7, 2012*

130. Wilkinson"Long range inspection and condition monitoring of rails using guided waves"*Proc. 12th Int. Conf. Exhib., Railway Eng., 2013*

131. R. Willsch, W. Ecke and G. Schwotzer"Spectrally encoded optical fibre sensor systems and their application in process control, environmental and structural monitoring"*Proc. SPIE Opt. Fibers, Appl., vol. 5952, 2005*

132. M. Wischke, M. Masur, M. Kroner and P. Woias"Vibration harvesting in traffic tunnels to power wireless sensor nodes"*Smart Mater. Struct., vol. 20, no. 8, p. 085014, 2011*

133. P. Wolfs, S. Bleakley, S. T. Senini and P. Thomas"An autonomous, low cost, distributed method for observing vehicle track interactions"*Proc. IEEE/ASME Joint Rail Conf., pp. 279-286, 2006*

134. R. Wong, R. Thomson and E. Choi"In situ pore pressure responses of native peat and soil under train load: A case study"*J. Geotech. Geoenviromental Eng., vol. 132, no. 10, pp. 1360-1369, 2006*

135. H. Yazdi"Intelligent condition monitoring of railway signalling equipment using simulation"*Proc. Inst. Elect. Eng.?Seminar Condition Monitoring Rail Transport Syst. (Ref. No. 1998/501), pp. 13-1-13-5, 1998*

136. Yick, B. Mukherjee and D. Ghosal"Wireless sensor network survey"*Comput. Netw., vol. 52, no. 12, pp. 2292-2330, 2008*

137. P. Yilmazer"The structural health condition monitoring of rail steel using acoustic emission techniques"*Proc. 51st Annu. Conf. NDT, pp. 1-12, 2012*

138. Zhou, M. D. Duta, M. P. Henry, S. Baker and C. Burton"Remote condition monitoring for railway point machine"*Proc. ASME/IEEE Joint Railroad Conf., pp. 103-108, 2002*

139. H. Zimmermann"OSI reference model-the ISO model of architecture for open systems interconnection"*IEEE Trans Commun., vol. 28, no. 4, pp. 425-432, 1980*

140. S. Z?ller"Efficient real-time monitoring of multimodal transports with wireless sensor networks"*Proc. 36th IEEE Conf. Local Comput. Netw., 2011*

Chapter 10

Challenges Toward Wireless Communications for High-Speed Railway

Bo Ai, Xiang Cheng, Thomas Kürner, Zhang-Dui Zhong, Ke Guan, Rui-Si He, Lei Xiong, David W. Matolak, David G. Michelson, and Cesar Briso-Rodriguez

ABSTRACT

High-speed railway (HSR) brings convenience to peoples' lives and is generally considered as one of the most sustainable developments for ground transportation. One of the important parts of HSR construction is the signaling system, which is also called the "operation control system," where wireless communications play a key role in the transmission of train control data. We discuss in detail the main differences in scientific research for wireless communications between the HSR operation scenarios and the conventional public land mobile scenarios. The latest research progress in wireless channel modeling in viaducts, cuttings, and tunnels scenarios are discussed. The characteristics of nonstationary channel and the line-of-sight (LOS) sparse and LOS multiple-input-multiple-output channels, which are the typical channels in HSR scenarios, are analyzed. Some novel concepts such as composite transportation and key challenging techniques such as train-to-train communication, vacuum maglev train techniques, the security for HSR, and the fifth-generation wireless communications related techniques for future HSR development for safer, more comfortable, and more secure HSR operation are also discussed.

KEYWORDS

Channel modeling, composite transportation, cuttings and tunnels, high-speed railway (HSR), operation control system, train control data, viaducts.

1. INTRODUCTION

With the fast development of high-speed railway (HSR) in the world, the maximum train moving speed has reached nearly 575 km/h, trialed by the French National Rail Corporation. With increased train speed, the issue of train operation safety has attracted more and more attention. The ground infrastructure (e.g., trackside equipment), the moving body (e.g., train), and the signaling system are in general the three main parts contributing to the HSR operation safety. Of these three parts, the signaling system, which is also called the train operation control system, is the key part and is regarded as the nerve center of the HSR system. A standard has been set up for the train operation control system, which is known as the European train control system (ETCS) for HSR and the communication-based train control (CBTC) system for subway [1]. To realize a safe operation of the HSR, a large number of studies on intelligent transport system applications have been conducted[2]–[3] [4] [5][6]. In order to make the train operation control system work better, maintaining a reliable communication link between the train and the ground, dedicated mobile communications play a key role. Such a communications system is called the global system for mobile communications for railway (GSM-R), a narrow-band communication system.

With the rapid growth of railway services, broadband communication systems for railway called long-term evolution for railway (LTE-R) will be deployed. Except for the train control data transmission, LTE-R is also expected to offer passengers services such as Internet access and high-quality voice or mobile video broadcasting. For instance, many wireless system deployments are currently being developed to realize driverless railway and subway systems, such as in New York, NY, USA; Paris, France; Malaga, Spain; and Singapore [7]. Another example of broadband communication application for railway is the radio communication systems that are devoted to providing Internet access to the train passengers. This is not a very common feature (as of today) because it is expensive to implement. However, some HSR operators begin to offer this service using ad hoc networks. On the other hand, telecommunications companies may extend its coverage area to railway tracks, permitting the passengers to access to it. This is very common in both subway and commuter trains. However, it cannot be considered as railway communications. One of the European HSR operators started in 2005 to offer Internet access to its customers [8]. The access technology was a combination of satellite access and terrestrial access based on IEEE 802.11[9]. There are many other technologies (in use or in the development stage) devoted to provide passengers with access to the Internet, such as the Mobile Wideband Global Link System [10]. In Spain, the TECRAIL project [11] has been carried out to evaluate the feasibility and performance of a broadband system (LTE) on railway communications and signaling systems. LTE for railway was determined in the 7th World Congress on High-Speed Rail to be the next-generation communication system for HSR [12]. Within IEEE 802.15 Task Group 4p, a standard for "Positive Train Control" is currently being developed [13]. The development of this standard has been triggered by the "Rail Safety Improvement Act of 2008" of the United States Congress. The scope of this standard is to monitor and control train movements. Various frequency bands in the range from 161 up to 5800 MHz are considered.

Whether it is a GSM-R, an LTE-R, or an IEEE 802.15.4p system, to guarantee a reliable communication, the major prerequisite condition is a thorough knowledge of the propagation characteristics of the wireless channel. Wireless channel modeling is the important basis and the essential means for communication network planning and optimization, the transmitter and receiver design, and the physical and upper layer key technologies selection. There are some influential research groups focusing on wireless channel measurement and modeling. Rappaport *et al.* are now working toward 60-GHz wireless communications and channel modeling [14]. Matolak *et al.* drew their attention on vehicle-to-vehicle (V2V) channel modeling [15]. As for the studies of channel modeling methodologies, Sivertsen*et al.*, Cheng *et al.*, and Guan *et al.* devoted themselves to the research studies on the multi-input–multi-output orthogonal frequency-division multiplexing (MIMO-OFDM) channel modeling[16], the geometry-based stochastic channel modeling [17], and the ray-tracing technique [18], respectively. Richter *et al.* proposed a novel flexible algorithm called RIMAX for channel parameter extraction from channel sounding measurements

[19]. In addition to the academic circles, some international academic organizations such as the European cooperation in the field of scientific and technical research (COST) and the WINNER project group [20] and some companies such as MEDAV in Germany and Elektrobit in Finland also do a lot of work on channel modeling, standardization, and the supply of channel measurement instruments. Although much remarkable progress has been achieved in the area of wireless channel modeling, most results are for public land mobile communications, but not for dedicated mobile communications for railways. As some research has shown, when key techniques developed for public land mobile radio (such as synchronization[21]–[22][23] and channel estimation [24]) are applied to the existing wireless network of HSR, the communication quality is poor, with a high rate of dropped calls and low data rate observed[25]. Matolak and Athens also found that nonstationary V2V channel models yielded results (for OFDM access) that are much different than those on stationary channels [26]. Following the conclusions drawn from [27] and [28], in a suburban area but with HSR-specific topography of cuttings [27] or viaducts [28], the path loss prediction results between the Hata model [29] and that of the actual test results obtained from Zheng-Xi HSR lines in China show that the conventional Hata model prediction error may reach 20 dB. What causes such large prediction performance degradation? The description in Sections II and III aims to provide an answer to this question.

Except for the fundamental problems of wireless channel modeling, the working frequency band should be taken into careful consideration due to serious interference and security problem. The next-generation broadband wireless communication system for HSR and other related technologies should be also mentioned, as will be described in subsequent sections.

2. DIFFERENCES BETWEEN RAIL MOBILE COMMUNICATIONS AND PUBLIC LANDMOBILE COMMUNICATIONS

Some of the typical features for HSR must be addressed. Special propagation scenarios, high moving speeds over 300 km/h, higher requirements of quality of service (QoS), harsh electromagnetic environments, and interference are the main differences between rail mobile communications and public land mobile communications.

- As is known, knowledge of the wireless channel is the fundamental basis for the planning of wireless communications network and the design of transceiver. A wireless channel has a close relationship with physical conditions, topography, and the surrounding environments. The typical railway environments are characterized by special scenarios such as viaducts, cuttings, tunnels, crossing bridges, railway stations, marshaling stations, and some combined scenarios such as the tunnels group and the combination of viaducts and tunnels. Based on the actual test data collected from Zheng-Xi HSR lines, conclusions are drawn that these special scenarios sometimes cause deep fading of the signals[30], [31]. In addition, obstacles such as passing trains, arched grids, wind barriers, and the acoustic barriers along the railway tracks may cause extra path loss due to obstruction, reflection, diffraction, or scattering. This may be cast into the concept of urban furniture proposed by Gougeon *et al.*[32], where the trees, and the wire rods, and the traffic signs, the street lamps in the urban areas are referred to as the urban furniture, which may cause the reflection or diffuse scattering loss. Similarly, the special scenarios along the railway lines can be regarded as the railway furniture.

- High speeds over 300 km/h for HSR bring new problems that may not be encountered in highway traffic. In general, for highway traffic with moving speeds below 200 km/h, the wireless channel is often assumed to be wide-sense stationary uncorrelated scattering (WSSUS), which may not pertain for HSR. For one reason, when the speeds of the train are over 300 km/h, the wireless channel exhibits rapidly time-varying and nonstationary features. For another reason, the line-of-sight (LOS) component exists in the majority of the HSR scenarios, and the number of scatterers and obstacles is limited. Thus, the spatial correlation between different multipath elements can be large, resulting in the correlated scattering. Simulations have been carried out based on LTE Release 9 V9.2.0 [33] to verify fast variation of the wireless channel within one subframe of 1-ms duration.

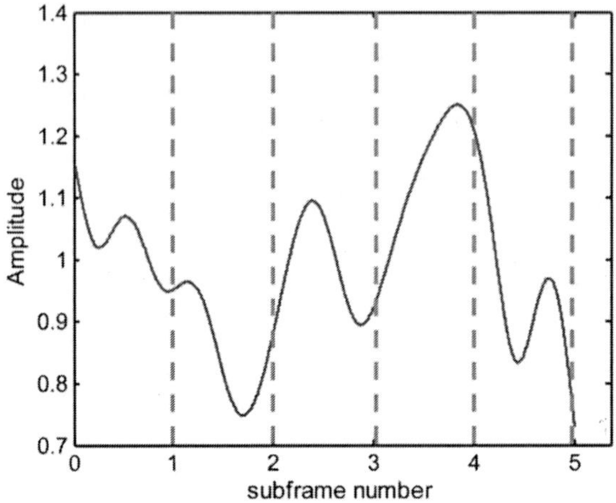

(a)

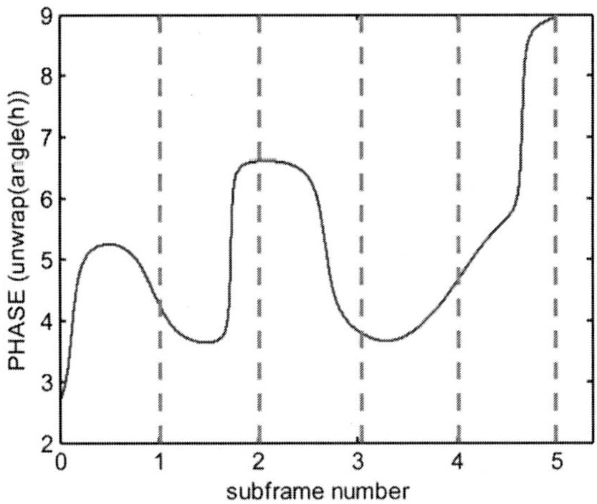

(b)

Figure 1. (a) Amplitude variation versus LTE subframe number. Data stream follows the LTE Release 9 V9.2.0 standard [32]. The sampling rate is 30.72 MHz, with 30 720 samples for one subframe; the working frequency is 2.4 GHz. Typical extended pedestrian A channel model. The moving speeds are 360 km/h. Five subframes are observed. (b) Phase variation versus LTE subframe number. Data stream follows the LTE Release 9 V9.2.0 standard [32]. The sampling rate is 30.72 MHz, with 30 720 samples for one subframe; the working frequency is 2.4 GHz. Typical extended pedestrian A channel model. The moving speeds are 360 km/h. Five subframes are observed.

Fig. 1(a) and (b), respectively, shows the amplitude and the phase variation of the signals versus LTE subframe number through the wireless channel. It is clearly shown in Fig. 1(a) and (b) that the channel varies quickly and randomly within one subframe. It is inappropriate to adopt the channel estimation method such as the linear interpolation method to estimate a rapidly varying channel since this method assumes the channel to be essentially constant over a subframe. This in another way to illustrate that the pilot design pattern presented in LTE Release 9 V9.2.0 cannot support high moving speeds. High mobility also introduces the high Doppler frequency shifts and spreads. If the train frequently moves between different physical environments, such as from viaducts to cuttings or tunnels, fast variation of the multipath structure results. This may introduce time-varying Doppler shifts superimposed on each multipath component, greatly increasing the difficulty of channel estimation and Doppler shifts estimation. The phenomenon of birth–death of multipath may arise. Moreover, frequent handover may happen when there is a high-speed train (HST) operation, which results in the drop calls or, more seriously, the interruption of reliable transmission of the train control signals.

1 The QoS requirements either in narrow-band communication system GSM-R or in broadband communication system LTE-R are somewhat higher than that of the public land mobile communication network, for example, GSM or LTE. Some performance evaluation indexes such as transmission interruption rate have never appeared in the GSM system. GSM is mainly used for voice communication, and it even allows for dropped calls occasionally. However, GSM-R is in charge of the train control signals transmission; hence, it requires much more reliable communication. In the standard of ETCS[1], the QoS requirement for GSM-R should be satisfied with testability, controllability, reliability, effectiveness, maintainability, safety, and security. Note that dedicated mobile communications in HSR are for train operation control services. Taking the parameter of transmission delay for train control signals transmission as an example, correct train control signals should be transmitted within 10 s for HSR operation control system and within 3.6 s for subway operation control system. If the operation control system received no information or incorrect train control signals in a required time range, it will start the automatic train control mechanism, and the train or the subway may automatically stop. For another example, the call setup time for trunking communication services in LTE-R is less than 300 ms, whereas such requirement is 5 s in LTE system.

2 GSM-R in China uses the 883- to 889-MHz frequency band for the uplink and the 930- to 934-MHz frequency band for the downlink, as shown in Fig. 2. This frequency band is shared with services offered by the operator of China Mobile. This can yield serious cochannel interference. The bandwidth of 4 MHz is divided into 19 frequency channels (from No. 999 to No. 1017). In a GSM-R network, the frequency reuse factor is from four to seven, which means that four frequency channels are available in each cell at most. The GSM-R system is required to support more and more services, such as train control, dispatching, and train and environment monitoring. Thus, the GSM-R frequency band will become more and more crowded, and the intrasystem interference is serious. WiFi technology is used in subway for train-to-ground communications in China. It operates at 2.4 GHz, an unlicensed frequency band, which will have serious interference and security problems. Moreover, there will be strong electromagnetic interference caused by the arched grid network used as train power supplies.

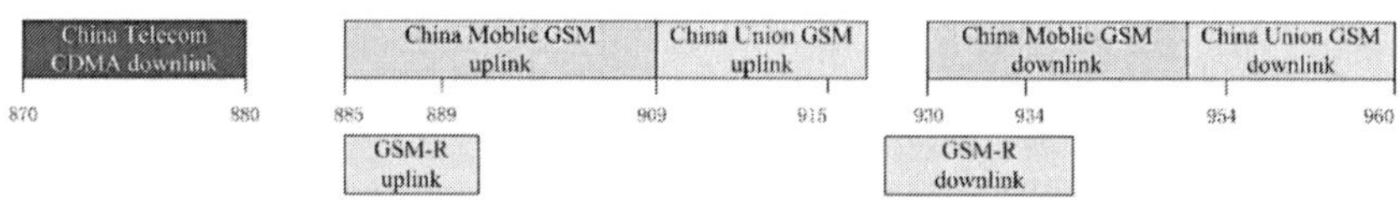

Figure 2. Diagram of 800- to 900-MHz frequency allocation in China.

The previously mentioned explanations of the main differences between the rail mobile communications and the public land mobile communications will be further addressed by a review of the latest research on channel modeling under HSR scenarios.

3. WIRELESS CHANNEL MODELING UNDERHSR SCENARIOS

Here, the latest progress in wireless channel modeling for HSR is presented. The scene partitioning for HSR, channel measurements, channel modeling for cuttings, viaducts and tunnels, and antennas modeling are discussed in detail.

3.1. Scene Partitioning for HSR

The path loss and multipath effects are in disparity under various propagation environments, leading to the impossibility of accurate pass loss predictions under different propagation environments with the same channel model. That is to say, radio wave propagation scene partitioning plays an important role in wireless channel modeling. Special HSR scenarios such as cuttings, viaducts, tunnels, and crossing bridges have significant impacts on the propagation characteristics. However, these HSR scenarios have rarely been investigated. Few channel measurements and standard channel models have actually been developed.

Several international organizations and related standards should be mentioned when we refer to the scene partitioning. International Mobile Telecom System 2000 defines nine terrestrial scenarios and four satellite scenarios [34]. The Universal Mobile Telecommunications System (UMTS) developed by 3GPP defines a smaller set of radio propagation environments, which adequately span the overall range of possible environments [35]. For practical reasons, these operating environments are an appropriate subset of the UMTS operating environments described in Recommendation ITU-R M.1034 [34]. The WINNER project group defines four typical scenarios, including in and around building, hot spot area, metropolitan, and rural scenarios. The only scenario appropriate for HSR defined in the WINNER project is the WINNER D2 model (rural moving network) [20]. However, this model has only one LOS path, inappropriate for most special HSR scenarios such as cuttings, tunnels, and crossing bridges. Moreover, the working frequency of WINNER D2a is at the 2- to 6-GHz frequency band, unsuitable for dedicated mobile communications network for HSR operating at 930 MHz.

As part of the latest work of the research group of Prof. Bo Ai, the HSR scenes are partitioned into 12 types. Detailed description about scene partitioning for HSR can be found in the literature [36]. Table I summarized the scene partitioning for HSR.

3.2. Channel Measurements at High Moving Speeds

Pass loss prediction modeling has a close relationship with the topography and the physical environments, whereas the multipath fast fading modeling has a close relationship with the moving speeds. High moving speeds make the work of the channel measurements more challenging. One scheme of channel measurements is the utilization of the existing GSM-R network. In order to pick up samples that can capture the fast fading characteristics, the distance between two adjacent fades should be, at most, corresponding to a time interval of, where is 100 m/s when the train moves at the speeds of 360 km/h. Then, the sampling interval between two adjacent samples at 900 MHz should be which means 1600 samples per second. In this scheme, the measurement instrument is equipped on the train, receiving signals transmitted from the running communications network. The cells for HSR communications provide zonal coverage along the railway tracks and generally use five or seven different frequency channels for cell multiplexing. Therefore, if we want to pick up samples at the same frequency channel when the train moves along the tracks, it should have to pass through other four frequency channels, which means that the actual sampling rate should be . Thus, this measurement scheme invokes strict requirement for the fast enough sampling rate of the measurement instruments.

Another scheme of channel measurement is the utilization of channel sounder, as shown in Fig. 3 [37].

This measurement is conducted on Beijing–Tianjin HSR. The No. 0 high speed integrated inspection train (NHSIIT) at the speeds of approximately 240 km/h is employed. The transmitter antenna is equipped on the NHSIIT, whereas a large crane is adopted to equip the receiver antenna. The viaducts and roadbeds are covered with agricultural fields with no scatterers and reflectors around. The campaign is performed with Propsound channel sounder at a center frequency of 2.35 GHz with 10-MHz bandwidth. The transmit power is 37 dBm. Direct sequence spread spectrum signals are used to probe the channel. The receiver records the samples on a hard disk as raw in-phase or quadrature data. However, limited by the transmit power of the channel sounder, the measuring range has to be limited within about 2 km. In addition, this scheme cannot measure different communication links simultaneously.

It is noteworthy that both approaches fail to have a good tradeoff between reflecting the practical propagation channel characteristics and presenting a comprehensive description of the channel parameters. The channel measurement, particularly the MIMO channel measurement at high moving speeds, still remains to be an uneasy task. This is probably why the WINNER group fails to come to a definite conclusion on high Doppler shifts and no standard channel models for HSR at high moving speeds have been developed yet.

Table 1 Scene Partitioning for HSR

Scenes	Definitions	Scenes	Definitions	Scenes	Definitions
S1	Viaduct	S5	Water-5a: River and lake areas	S9	Mountain-9a: Normal mountain
			Water-5b: Sea area		Mountain-9b: Far-mountain
S2	Cutting	S6	Urban	S10	Desert
S3	Tunnel	S7	Suburban	S11	Combination scenario-11a: Tunnel-group
					Combination scenario-11b: Cutting-group
S4	Station-4a: Medium or small sized station	S8	Rural	S12	In-carriage-12a: Relay transmission
	Station-4b: Large station				
	Station-4c: Marshalling station and container depot				In-carriage-12b: Direct transmission

3.3. Channel Modeling Under Viaduct Scenario

Viaduct is the most common scenario for HSR. Viaducts, on average, make up 40% for a typical railway line [38], [39]. Fig. 4 shows an actual viaduct scenario in HSR, where the railway track is usually placed on the surface of a viaduct, to ensure flatness of the tracks and to avoid steep inclines, thus enabling high speeds of the train. As reported in[38]–[39][40], a typical HSR viaduct has a height from 10 to 30 m. The base station (BS) antennas are usually positioned 10–20 m away from the railway tracks with a relative antenna height (vertical distance between the BS antenna and the receiver antenna) of 20–30 m. The receiver antennas are placed in the middle part of the train, mounted on the top at a height of 30 cm above the roof of the train.

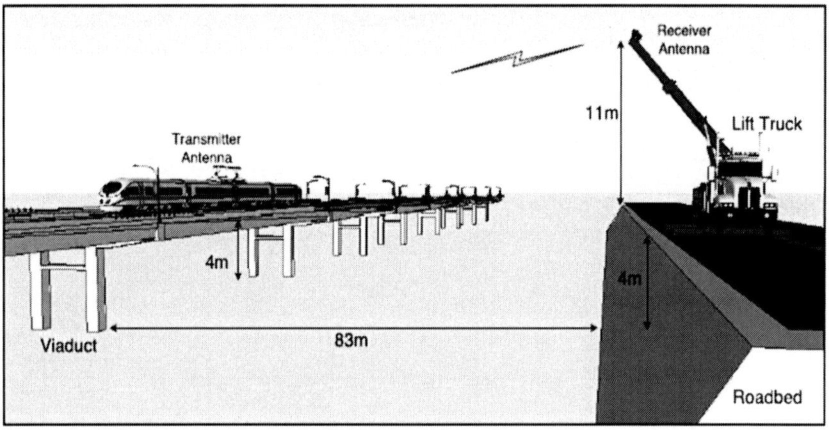

Figure 3. Channel-sounder-based channel measurements in high-speed moving train [37].

The propagation channel of the viaducts is found to be greatly influenced by the physical environments, the directional BS antennas, and the viaduct height. In[41] and [42], a special bottom area of the directional BS antenna in HSR is observed and defined, leading to a series of breakpoint models in the viaduct channels, e.g., the path loss is found to be a piecewise function of the distance with a break point at 400 m. Moreover, the path loss is also very sensitive to the viaduct height. In[40], it is found that a higher viaduct leads to a larger path loss exponent , and a simplified model is proposed, which is expressed as

$$n = k_1 H + k_2 h + \frac{k_3}{h} + k_4 \tag{1}$$

where H and are the viaduct height and the relative antenna height, respectively. The coefficients can be obtained by using a regression fit. Equation (1) shows a very interesting phenomenon that, with the increase of the viaduct heights, the path loss exponents increase as well, different from our traditional ideas and observations. This is because a higher viaduct leads to fewer reflected and scattered components, with the reduced power at the receiver compared with that of the plain propagation. Following these observations in the measurements, He *et al.* proposed a path loss model covering both height dependence and distance dependence for the HSR viaduct scenarios in [38]. An example plot of the path loss model, together with the measurements, is shown in Fig. 5, where the measurements were conducted on three 10-m-high viaducts of HSR line at 930 MHz. We can see that the proposed piecewise model in [38] and the measurements are in good agreement. The directional BS antenna leads to a breakpoint of the distance-dependent path loss at 400 m, and the curve of the proposed path loss model fits the measurements well. The Hata model [29] and the WINNER model in D2 scenario [20] path loss curves are also plotted for comparison, where we can generally observe 10- to 30-dB errors, indicating poor fit to the actual measurements.

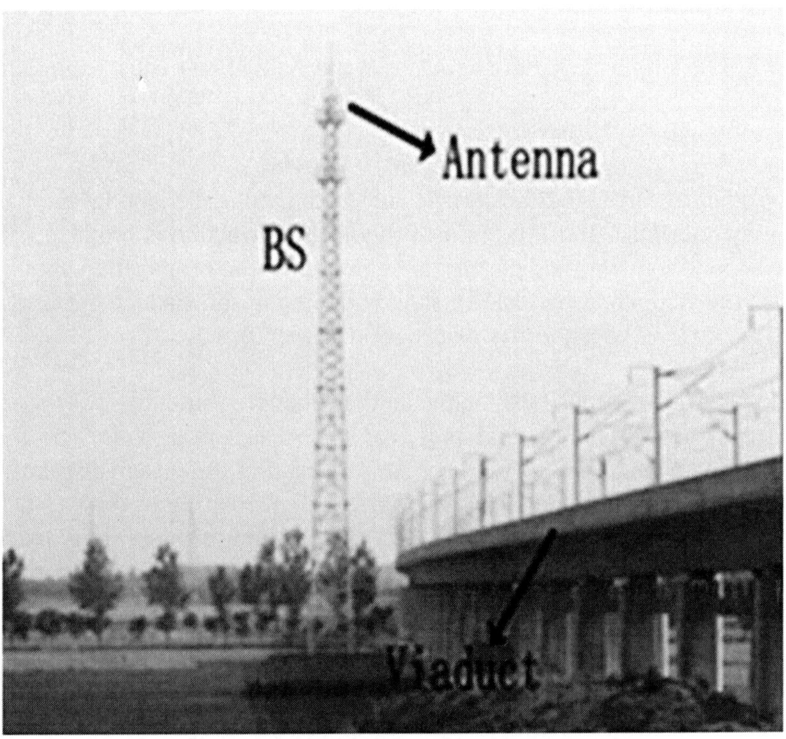

Figure 4. View of the viaduct scenario in HSR.

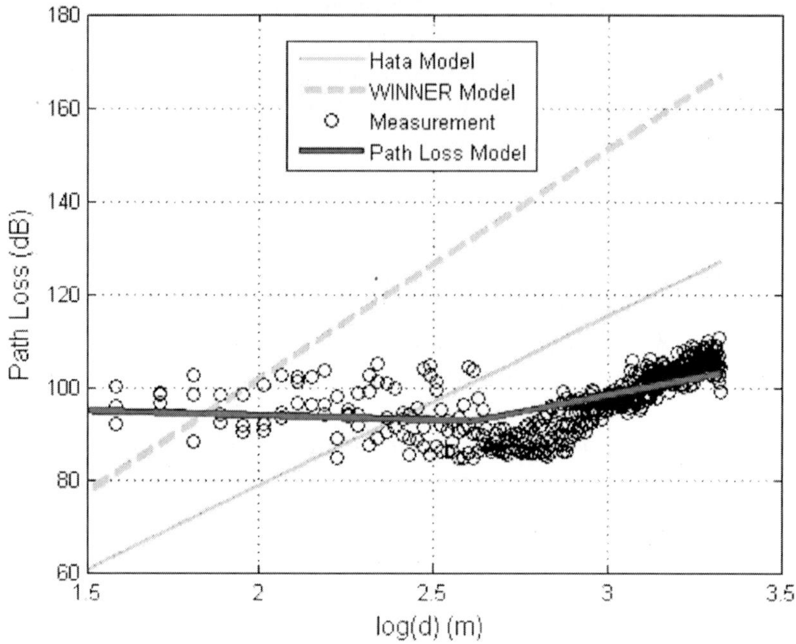

Figure. 5. Example plot of the path loss in the viaduct scenarios, from [38].

It is also noteworthy that a heuristic analytical propagation model using the geometrical optics and the uniform theory of diffraction is proposed [43] and is found to have good performance to predict the propagation pass loss under the viaduct scenarios. The shadow fading in the viaduct scenarios is found to follow the lognormal distribution [38], [40].

As for the small-scale fast fading in the viaducts, both the root-mean-square error distribution test [44] and the Akaike information criterion (AIC)-based evaluation [38] are used. It is found that the Ricean distribution outperforms the test and evaluation results. Similarly, the Ricean -factor is also found to be a piecewise function of the distance due to the directional BS antennas in HSR. A detailed formulation is presented in [38] covering the effect of the viaduct height on channel characteristics. Moreover, the composite channel in the viaduct scenarios is analyzed and is found to follow a Suzuki distribution. As was reported in [38], the viaduct reduces the severity of fading by providing clear LOS propagation with few received reflection and scattering components. A higher viaduct makes the reduction of the severity of fading more distinct.

3.4. Channel Modeling Under Cutting Scenario

Cutting is another common scenario in HSR. It is used on uneven ground, to help the HST pass or "cut" through large obstacles such as hills [42], as shown in Fig. 6. It ensures flatness of the railway lines. The regular cutting is very common today, where the steep walls on both sides of the rails have almost the same depths and slopes [43].

Figure 6. View of the cutting scenario in HSR.

The cutting has been divided into deep cuttings and shallow cuttings, and the typical structural parameters of ten groups of the cuttings in HSR are shown in Table II [46].

In Table II, and are the widths of the top and bottom of the cutting, respectively. The widths of the cutting and the depth are found to be influential in radio wave propagation. In [45], a set of statistical modeling methods based on the goodness-of-fit evaluation is proposed to examine the effect of the structure parameters of the cuttings on the propagation models. In [46], the path loss models for the cuttings with different depths are provided. However, systematic investigations of the height dependence with sufficient number of samples to give statistically relevant results are still a task for future research.

Table 2. Parameters for Cuttings

	Deep Cuttings							Shallow Cuttings		
Number	**1**	**2**	**3**	**4**	**5**	**6**	**7**	**8**	**9**	**10**
W_{up}(m)	52.01	50.86	55.26	55.72	53.93	58.30	63.17	48.38	49.77	50.24
W_{down}(m)	18.77	16.85	18.57	18.25	14.78	15.16	18.51	17.14	17.97	16.96

An important feature of the cuttings is the rich reflection and scattering components caused by the steep walls on both sides of the railway tracks. Abrishamkar and Irvine regarded the cutting as an open area with low reflection [47]. However, based on the actual test data and theoretical analysis, He*et al.* proved that the conclusion is incorrect [45]. It is found that, although the high BS antennas create a relatively clear LOS channel, the steep walls on both sides introduce rich reflection and scattering components at the receiver. By using the AIC-based evaluation, the percentages of the best fit for four types of distributions are compared, as shown in Fig. 7. The Ricean distribution has the best fit percentage of 68.39%. In the cutting scenarios, it is found to have a very low -factor around 1.52 dB due to the rich scattering components [30].

3.5. Channel Modeling Under Tunnel Scenario

Although many literature deal with the channel modeling inside tunnels, there are still two main problems unsolved. The first is the multimode waveguide mechanism, and the second is the breakpoint determination for different propagation mechanisms inside tunnels. Latest research has found many interesting phenomena inside the tunnels. One is the newly proposed concept of *far line of sight (FLOS)* in[48]. Mostly, the direct ray makes the main contribution to the received signal, with reflections contributing much smaller amounts. However, in some special environments such as the subway tunnel, with the increase of the distance between the transmitter and the receiver, the influence of the direct ray is becoming negligible, whereas the influence of the reflection is becoming considerable, leading to a special characteristic of delay distribution. The FLOS region is usually located 100–300 m apart from the transmitter inside tunnels.

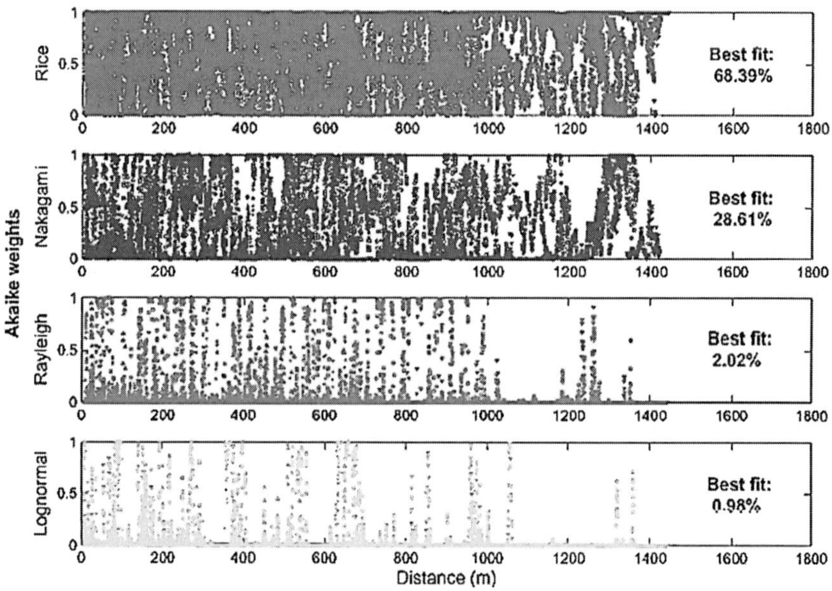

Figure 7. Akaike weights and the percentage of the best fit for four candidate distributions based on the measurements in cuttings, from [45].

Recently, it has been also found that the delay characteristic inside the subway tunnel is affected by the presence of the station, which changes the regular geometric structure of the tunnel, leading to extra reflection components and an increase in the delay spread [49]. In Fig. 8, we present an example plot of the measured power delay profile (PDP) in the subway tunnels of Madrid [49]. The parameters of the classical Saleh–Valenzuela model were derived to characterize the long delay clusters in the subway tunnels. Fig. 8 presents the measured PDPs at 2.4 GHz for LOS, non line of sight (NLOS), and FLOS regions in a 2600-m-long slightly curved subway tunnel. It shows that, in the subway tunnel, the 1- to 3-   delay clusters (from the reflections caused by the station in the subway tunnel) are very strong, even in the LOS and FLOS regions. The measured large delay spread shows that the subway tunnel has large delay dispersion when the train is close to the station, which should be carefully considered when designing the subway communication networks. This observation is useful for the dedicated mobile communication network design within HSR tunnels, since there are also many railway constructions around the entrance/exit of the HSR tunnels, for example, the power lines and the pillars. These scatterers may lead to a large delay spread around several microseconds.

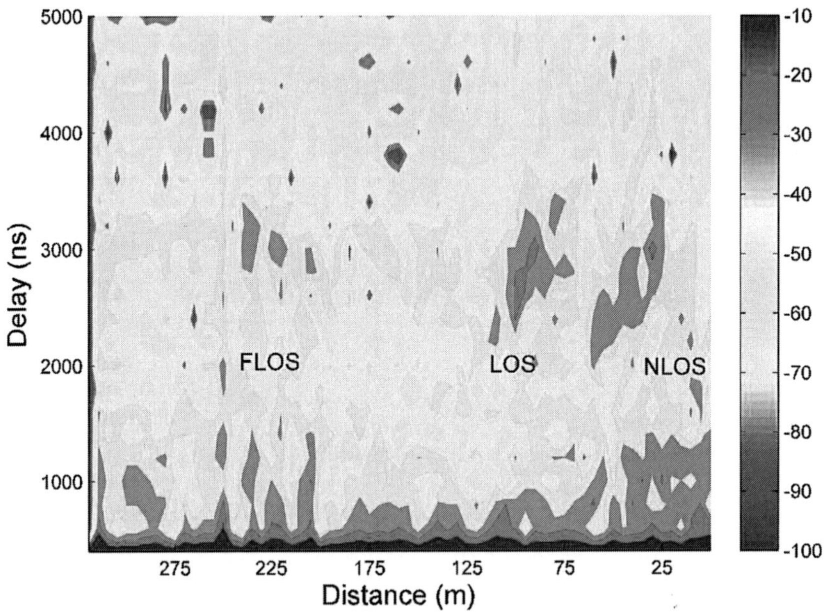

Figure 8. Example plot of the PDP measured inside a subway tunnel in Madrid, Spain, from [50].

Another finding is the *near-region* phenomena inside tunnels based on the referenced antenna configuration on the train in a Madrid subway [50].

Along with the increase of the distance between the transmitter and the receiver, for the relatively small-size users (such as pedestrians and light cars), the free-space propagation mechanism appears first at the shortest distance, the multimode waveguide effect establishes later at slightly larger distance. For relatively large-size users (such as long trains in subway tunnel), the first appeared mechanism is the *near shadowing*. The corresponding definition and critical condition about near shadowing are presented in [51]. Then, the free-space propagation occurs before the waveguide propagation if LOS recovering is earlier than the end of the free-space propagation. Otherwise, the multimode waveguide propagation is behind the near shadowing effect; no free-space propagation exists.

Fig. 9 shows the near-region phenomenon inside tunnels. In order to present a clear explanation for Fig. 9, some introductions are given as follows.

- The channel path loss in the *free-space propagation zone* can be modeled by the free-space model.
- The propagation loss in the near shadowing zone is modeled with the utilization of the principle of least squares curve fitting on the measured data in [52]. Details of the modeling and the related parameters can be found in [51] and [52].
- The critical condition of the near shadowing phenomenon can be defined as the case that the widest part of the 60% of the first *Fresnel zone* is touched by the vehicle.
- In the *multi*mode propagation zone, a limited multimode model is utilized. By assuming the arched and circular tunnel approximated to an equivalent rectangular tunnel, the quantity and the type of modes are determined by the frequency and the structure of the tunnel. Then, by introducing two modified factors (the tilt loss and the roughness loss)[53], the propagation loss at all the coordinates can be analytically calculated. Details can be found in [54].
- Dividing Point 1 in Fig. 9 locates at the distance when the maximum first Fresnel zone touches any one of the tunnel walls. Details and simplified formulas of Dividing Point 1 in rectangular, circular, arched, and arbitrary cross-sectional tunnels are given in[55]–[56] [57][58]. This model has been validated by five groups of measurement campaigns conducted in various tunnels. Dividing Point 2 between the near

shadowing zone and the next zone locates at the distance of the length of the large-size user. By assuming an equivalent rectangular or a circular tunnel, the location of Dividing Point 3 is defined as the distance where the second-fundamental modes have suffered one reflection from the vertical or the horizontal walls.

In Fig. 9, the equations can be employed to realize fast communication network planning for various types of users. In order to keep good readability of this paper, the definitions and derivations of the equations are not presented here. However, details of all the cited equations and notations can be found in [59]. By making a brief review on the advancement of wireless communication systems, it is found that the near-region phenomenon inside tunnels lengthens considerably owing to the increase in working frequencies. Above all, almost all the channel models developed based on the measurements from one tunnel cannot be suitable for another one. There is a lack of commonly used channel models for tunnels.

Another special scenario for HSR, the crossing bridges are usually built over the cuttings to ensure necessary transportation for people around the HSR. The crossing bridge over the cutting leads to NLOS propagation within a short distance, causing an extra large-scale loss of about 5 dB [31]. However, there is no obvious change in the small-scale fading behavior caused by the crossing bridges [41]. For a high-speed moving train, this temporary change from LOS to NLOS will not significantly affect the fading behavior at the receiver.

3.6. Antennas Modeling

The antennas modeling should be independent with that of the wireless channel modeling. These are two different things. One is for the exploration of the wireless channel characteristics, and the other is for the antennas performance evaluation. Many people misunderstand because the test data received by the receiver are affected by both the antenna settings and the channel conditions. The aim of the antennas modeling is to establish the relationship between the condition of the field strength coverage and the variation of the height, the angle, and the polarization of the antennas. As for multiantenna system such as MIMO system[3], or the massive MIMO system [60], which is regarded as one of the key technologies in fifth-generation (5G) mobile communication systems, the relative position of different antennas and the degree of the correlation among different antennas are all contributing to the antennas modeling. Up to now, very few papers have dealt with the research of antenna array structure for HSRs.

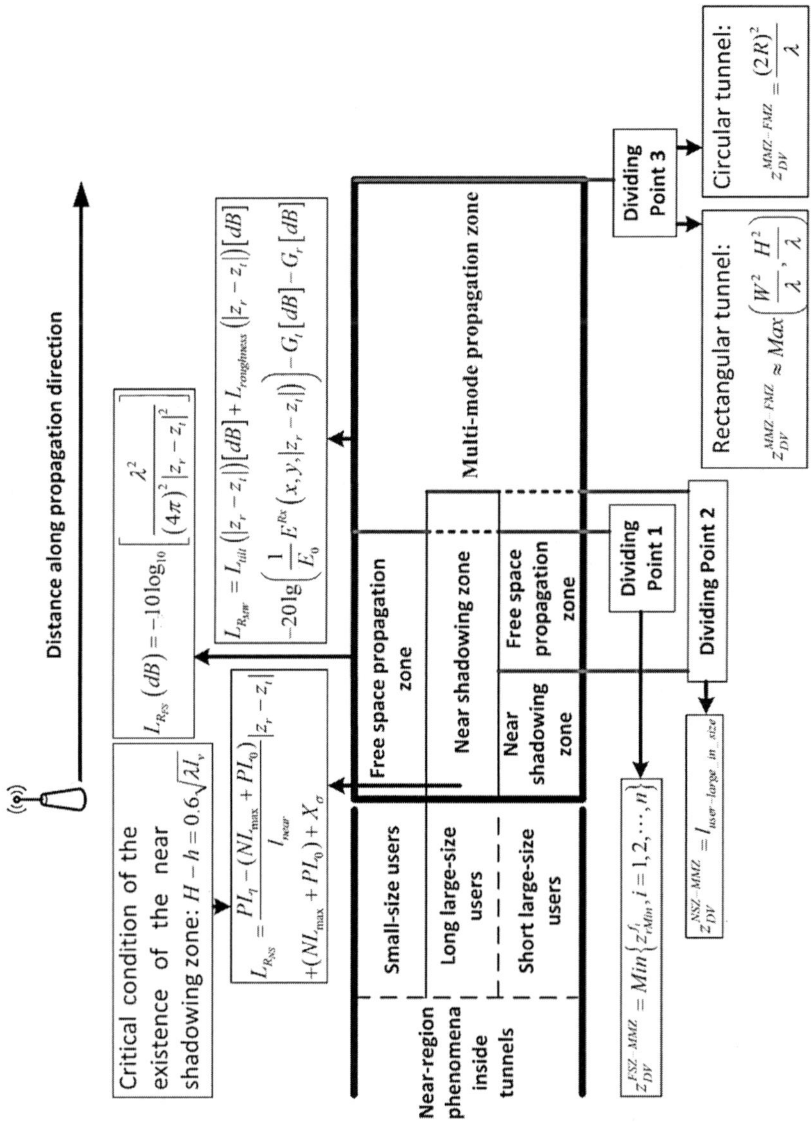

Figure. 9. Panorama of the propagation inside the near region of tunnels [51].

In [61], the authors conducted the research on the performance comparisons in terms of channel capacity when the antenna array has different forms, such as uniform linear array (ULA), uniform circular array (UCA), star, and uniform rectangular array (URA), as shown in Fig. 10.

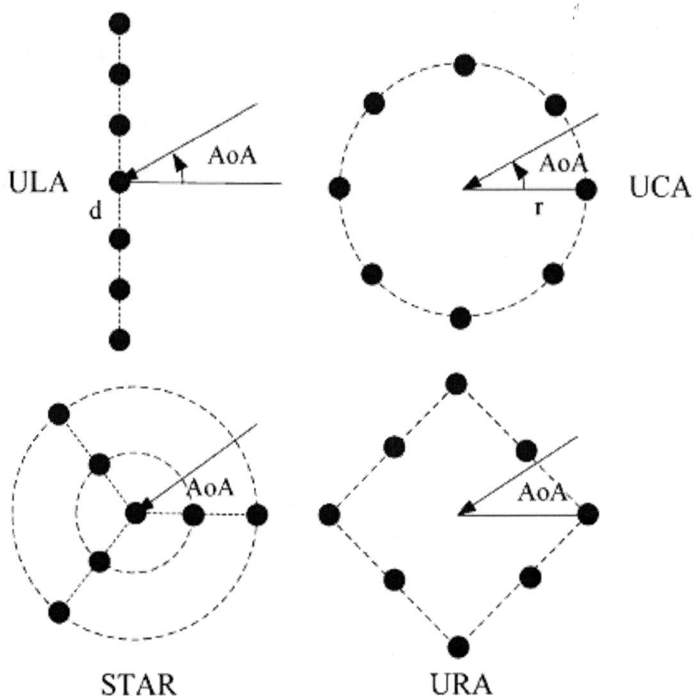

Figure 10. Different antenna array configurations, including ULA, UCA, star, and URA.

The ergodic channel capacity simulation results of four different antenna array configurations are shown in Fig. 11. The channel capacity of each antenna array is simulated according to the real HST scenario. The simulation parameters are as follows. It is assumed that the BS is located 30 m away from the track. The carrier frequency is 900 MHz. The velocity of the train is set to be 360 km/h. Thus, the maximum Doppler shift is 300 Hz. The number of antenna elements is eight, the same in ULA, UCA, and URA.

The star array has seven elements due to its symmetrical configuration. It is assumed that all these elements are vertically polarized and have an interelement spacing of , where is the wavelength of the light. The D2a scenario in the WINNER II channel model with LOS path has been adopted [20].

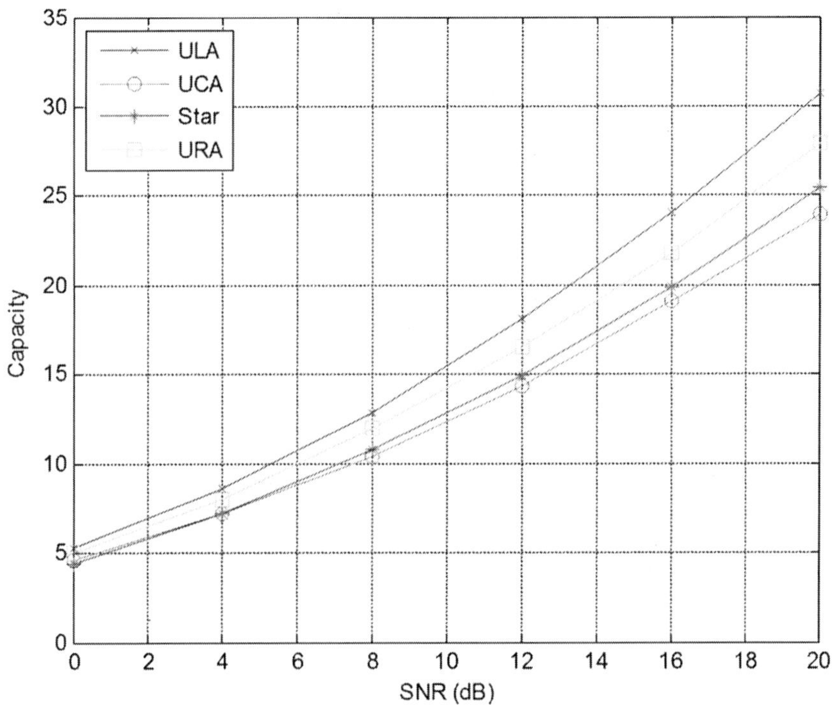

Figure 11. Capacity comparison of four antenna array configurations, including ULA, UCA, star, and URA.

From Fig. 11, we can see that the performance of ULA is superior to all the other three antenna array forms when the simulation parameters are the same. The performance of ULA is also related with the azimuthal orientation of the array. In the other three types of antenna arrays, URA provides a better performance.

Based on the wireless channel characteristics in HSR, there is often one dominant LOS component and sparse multipath in the environment, generally degrading the MIMO performance. An optimal MIMO antenna, which is called reconfigurable antenna array, is designed in [62] to maximize the channel capacity in the low-rank LOS Ricean channel. An optimal MIMO antenna combination for HSR is proposed in [63]. It contains four types of parameters for MIMO antenna array configuration corresponding to the four different channel regions defined in [61] when the train runs toward the BS. For the optimal MIMO antenna and location, Christopher *et al.* [64] have carried out a vehicular (automobile) MIMO measurement to investigate the effect of antenna array location, roof, or two sides of the car on the MIMO system performance. The array on the roof can obtain more signal power. However, the diversity gain can be obtained if the array is located on the two sides of the car. A new tool, which is called "*Virtual Drive*," is developed by Reichardt*et al.* [65], [66]. It consists of the vehicular environment, a traffic model, and a 3-D ray-tracing tool. With this software tool, different V2V or vehicle-to-infrastructure (V2I) scenarios can be simulated, and the channel is calculated by *ray-tracing method* to find an optimal MIMO antenna configuration.

Up to now, there have not been so many scholars or research groups paying their attention on the MIMO or the massive MIMO channel modeling for HSR scenarios. It is still on its way.

3.7. Availability of GIS Data

Detailed geographic data have to be made available in order to model the different scenes. On a large scale, such data are available for terrain height only using data retrieved, for example, from the Shuttle Radar Topography Mission (SRTM) [67]. Other data can be found using the framework of the OpenStreetMap initiative [68]. For example, such OpenStreetMapData have been used to model car-to-X-communication in urban environments [69]. Current publications describing the modeling of objects such as viaducts, crossing bridges, or train stations are using data that have been manually measured by a laser distance meter, e.g., [31]. When applying the corresponding propagation models, an effort has to be made to derive all relevant data at least in a semiautomatic way. Geographic information system (GIS) data are the fundamental basis for ray-tracing method, which is used for the software simulation of the actual physical scenes and the accuracy verification of the wireless channel models.Moreover, GIS is also very important for Global Positioning System (GPS) location in HSR [70]. For example, GPS-based train positioning can be used to detect the missing train compartments left on the railway tracks. Based on the location information of the missing train compartments, the dispatcher and the train driver can obtain prewarning messages to avoid the collision of the trains.

4. CHALLENGING KEY TECHNOLOGIES FORHSR COMMUNICATIONS

In addition to the wireless channel modeling for HSR, some challenging key technologies for future development of HSR should be paid more attention.

4.1. Working Frequency Band

As pointed out in Section II, GSM-R shares the same working frequency band with the public land mobile communication network, as shown in Fig. 2, which will cause serious cochannel interference. With the increase of various railway services, the frequency resources will be seriously limited, particularly in railway hub areas. As for subways, some commonly unlicensed frequency bands such as 2.4 GHz used in CBTC systems may risk the integrity and confidentiality of the train control data.

With the emergence of the new services for HSR, the comfortable and humanized demands for future HSR development, broadband communications for HSR have come to the agenda. LTE system can achieve high capacity, low latency, and high reliability[71]. Twenty frequency bands have been defined by 3GPP for LTE frequency-division duplexing (FDD) and nine bands for LTE time-division duplexing[72]. IEEE 802.15 Task Group 4p has defined a total of 16 frequency bands [13], including four bands between 161 and 220 MHz, one band at 450 MHz, seven bands between 770 and 928 MHz, and five bands between 4965 and 5800 MHz. With regard to the suggestions of LTE frequency band allocation from 3GPP, the planning and the usage of the LTE frequency band in China are shown in Table III.

From Table III, it is known that 700 MHz is used by the State Administration of Radio Film and Television of China. It is easy to realize a wide range of communication network coverage with low cost of construction. However, such frequency band is influenced by the national policies and difficult to be released. The 1.4-GHz frequency band has been used as the municipal LTE communication trial network for the subway system in China. If 1.4 GHz is adopted for LTE-R dedicated mobile communication network for train-to-ground communications, there will be serious cochannel interference. The 1785- to 1805-MHz frequency band has been allocated for the dedicated communications such as the Fire Bureau and the sea ports in economically developed eastern coastal areas. However, it is also possible to be allocated to LTE-R due to less interference and easy application of the frequency bands. As we know, the higher the working frequency band is, the larger the path loss and the more substantial the signal fading. Higher frequencies exhibit weaker diffraction capability, which is very important for the railway dedicated mobile communications. If 2.6 GHz is adopted, with typical transmit power levels, the radius of a cell is no more than 500 m, resulting in a dense layout of the BSs for HSR. This will cause a large investment and frequent handovers.

Above all, it is suggested that the 800-MHz 1800-MHz frequency band dual mode is deployed for HSR dedicated mobile communications. The high priority service is working in the 800-MHz frequency band, and the low priority service is working in the 1800-MHz frequency band.

4.2. LTE for Railway

As GSM-R manufacturers said, the life cycle of GSM-R will be coming to an end in 2025, whereas the International Union of Railways (UIC) said that the development of the next-generation mobile communication systems for railway is to bypass third-generation mobile communications and come to LTE-R directly. As announced by both GSM-R manufacturers and UIC, however, LTE-R must be designed to be backward compatible with GSM-R. The IP packet network in LTE replaces the circuit domain data, whereas the voice services for railway are mainly supported by the circuit domain. Therefore, in the early stage of the construction of the LTE-R system, the packet data services and the multimedia services will be carried by the LTE-R network, whereas the voice services and the train control data services in the circuit domain will still be carried by the GSM-R network. With the gradual maturity of LTE-R, all the railway services will be then gradually transferred to the LTE-R system. Above all, the coexistence of LTE-R and GSM-R will be required for a long time.

Table 3. LTE Frequency Band Allocation in China

Frequency band	3GPP suggestion	Services
1850-1900 MHz	TDD/FDD (Uplink)	1800-1920 MHz: TD-SCDMA; 1850-1880 MHz: mobile and fixed communications
1880-1920 MHz	TDD	TD-SCDMA, trail communication network for TD-LTE
1910-1930 MHz	TDD	CDMA2000
1930-1990 MHz	TDD	1920-1935 MHz: CDMA2000; 1940-1955 MHz: WCDMA; 1960-1990 MHz: mobile and fixed communications
2010-2025 MHz	TDD	TD-SCDMA
2300-2400 MHz	TDD	2320-2370 MHz: TD-SCDMA, trail communication network for TD-LTE; 2370-2400 MHz: mobile and fixed communications, radio positioning
2500-2690 MHz	TDD	TD-LTE of China Mobile
around 700 MHz	FDD	Mainly used by the State Administration of Radio Film and Television of China
around 800 MHz	FDD	806-821 MHz, 851-866 MHz: Digital trunking system; 825-840, 870-885 MHz: CDMA; 885-889 MHz, 930-934 MHz: GSM-R; 890-909 MHz, 935-954 MHz: China Mobile GSM900; 909-915 MHz, 954-960 MHz: China Unicom GSM900; 841-869 MHz bands are unused.
around 1.4 GHz	FDD	Mobile and fixed communications, satellite communication services
around 1.7 GHz	FDD	1710-1725 MHz, 1805-1820 MHz: China Mobile GSM1800; 1745-1755 MHz, 1840-1850 MHz: China Unicom GSM1800; 1785-1805 MHz: for dedicated mobile communications
1920-1980 MHz	FDD	1920-1935 MHz: CDMA2000; 1940-1955 MHz: WCDMA; 1960-1990 MHz: mobile and fixed communications

What is the need for the development of LTE-R system for railways? The E-train project group of UIC has summarized over 200 kinds of communication services for railways[73]. As the investigation results from the E-train project concluded, to ensure the safe operation of the trains, to elevate the operation efficiency, and to improve the QoS, there are over 200 kinds of services: train dispatching, train control, train operation communication, safe video monitoring, wireless monitoring for the railway infrastructure, monitoring of the train state, remote fault diagnosis, automatic train operation, mobile ticketing, travel information, transportation and the tracking management of containers and dangerous goods, etc. These railway services requirements appeal to the coming era of the railway broadband mobile communications.

LTE-R is not just a simple move from LTE for public land mobile communications network to the railway environments. It has some differences from public LTE in the aspects such as radio propagation environments, network layout, services, and QoS requirement. Based on LTE communication network architecture[74], the suggested architecture for LTE-R is shown in Fig. 12, consisting of only two user-plane nodes (EnodeB and S/P-GW), simpler than GSM-R. The handovers between EnodeBs are handled over the X2 interface rather than in the BS controller. There is a lack of maturity on time-division LTE (TD-LTE) in high moving speeds of HST. Therefore, the hybrid network mode of TD-LTE and LTE FDD for railways is required.

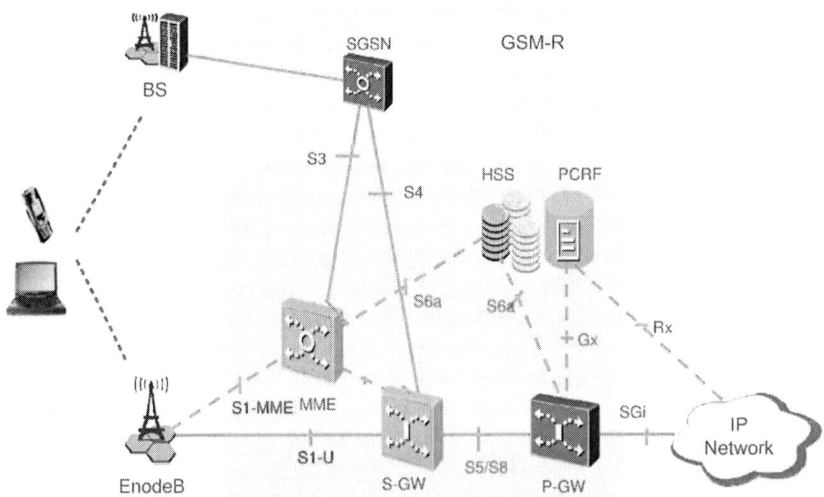

Figure 12. Communication network architecture of LTE-R.

Due to the high mobility and high requirements of QoS, the key technologies for LTE-R include the fast synchronization, the channel estimation and equalization, the Doppler shift estimation and correction, the wireless channel modeling, and the MIMO technique with strong correlation property. The frequency band used by the next-generation railway mobile communication network is higher than that by the GSM-R network, resulting in a much smaller distance between two adjacent BSs. Thus, frequent handover appears, and a fast handover technique with high reliability is needed.

Based on the special services requirements for railways and the typical characteristics of broadband communications, the preferred performance indexes for broadband communications of railways are given, although, up to now, there are no agencies or research teams giving such parameters. What we have listed in Table IV are for reference only.

Table 4 Recommended Parameters for LTE-R

Performance index	Recommended value	
Bandwidth	≥10 MHz	
Data rate	Uplink: 10 Mbps, Downlink: 4 Mbps	
Peak spectral efficiency	Medium and low moving speed (0-200 km/h)	Uplink: 2.5 bps/Hz, Downlink: 5 bps/Hz
	High moving speed (>200 km/h)	Uplink: 1.5 bps/Hz, Downlink: 2.5 bps/Hz
Handover success rate	≥ 99.5%	
Connection loss rate	≤ 10^{-2}/h	
Network attachment delay	≤ 30 s (95%), ≤ 35 s (99%), ≤ 40 s (100%),	

4.3. Nonstationary and LOS Sparse Channel

High mobility leads to the violation of WSSUS condition for wireless channel under HSR scenarios. Nonstationary channel is a very interesting issue with great value for the selection of physical layer technologies such as synchronization, channel estimation, and equalization. It has been mentioned in Section II that high-speed moving train will pass through many combined scenarios within a very short time interval. How to determine the boundaries of stationary channel regions and how to deal with such fast-varying channel for the receiver are very crucial and challenging. Birth and death process of multipath with Markov chains may be utilized to describe such phenomena.

With the development of intelligent transportation system [75], [76] and the LTE-R system, MIMO system under high-speed rails scenarios has becoming a hot topic. Under most HSR scenarios, the BSs are located less than 30 m away from the tracks, and most of the BS antenna heights are more than 30 m; hence, the LOS component exists in the majority of the HSR scenarios, and the scatterers are limited, revealing sparse channel characteristics. The spatial correlation between multiantennas elements is high owing to the linear dependence of the LOS rays' phases on the receive elements. Then, the concept of LOS MIMO appears. The effectiveness of using multiple antenna is demonstrated in LOS propagation environments [77]–[78][79]. The LOS MIMO channel under HSR scenarios has its own characteristics, which needs to be approached in a different way. It is randomly time varying: its identifiability from the measurements, the statistical characterization, and the modeling is rather challenging [80], [81]. In [82], single-bounce scattering is assumed for both fixed and moving scatterers. The high mobility of the transmitter, the receiver, and the surrounding vehicles results in time-variant angles of departure and angles of arrival (AOAs). This fact makes the model nonstationary. A nonstationary wideband geometry-based stochastic model (GBSM) for MIMO under HST channels is presented in[83]. In [84], the wideband LOS MIMO channel models under HSR scenarios were investigated, and the analysis was based on a training and communication scheme employing signaling over orthogonal short-time Fourier (STF) basis functions, which STF signaling naturally relates sparsity in delay Doppler to coherence in time and frequency. A novel wideband MIMO channel model, which is referred to as the*structured model*, based on the eigenvalue decomposition of the wideband channel correlation matrix is presented in[85]. A GBSM channel model based on the tapped delay line (TDL) for wideband MIMO mobile-to-mobile Ricean fading channels is proposed in[86], which has the ability to study the impact of the vehicular traffic density on channel statistics with different time delays.

An interesting issue for LOS MIMO is the maximization of the channel capacity. Under the HSR LOS MIMO scenarios, the channel is normally rank deficient owing to the correlated mutipath. To overcome this problem, the reconstruction of the antenna array can be employed, where the antenna elements are positioned to preserve orthogonality, hence maximizing the channel rank[87]–[88] [89] [90][91]. To investigate the characteristics of the LOS MIMO system, a suitable channel model is required. It is common for an LOS MIMO channel response matrix to be modeled as

$$H = \sqrt{\frac{K}{K+1}} H_{\mathrm{LOS}} + \sqrt{\frac{1}{K+1}} H_{\mathrm{NLOS}} \tag{2}$$

where denotes the LOS MIMO channel response matrix, modeled by the TDL model. is the channel coefficient corresponding to the free-space responses between all elements. is the NLOS channel coefficient due to reflection, diffraction, and scattering from the transmission environment. represents the Ricean -factor converted to linear scale. The validity of the maximum capacity criterion applied to realize the LOS MIMO channel is investigated for HSR scenarios[63]. In [62], a versatile methodology for maximizing LOS MIMO capacity at any operating signal-to-noise ratio is presented. Numerical results demonstrate that, by simply adjusting antenna spacing as a function of the distance, the orientation, and the spacing of the arrays, significant capacity gains are achievable.

4.4. T2T Communication

Train-to-train (T2T) communication is proposed as another auxiliary safety-guaranteed measure for HSR communication. It is first presented by Garcia *et al.*[92]. This idea was originally from the automatic identification system, the traffic alert and collision avoidance system, and the car-to-car communication system. In[92], the authors described an overview of the state of the art in railway collision avoidance for maritime transportation, aircraft, and road transportation, proposing a channel model for direct T2T communication appropriate for 400 MHz. Although the presented system has gained some progress in physical layer design, it only supports the train operation velocity lower than 200 km/h at 400 MHz, which is not suitable for the HSR with the moving speeds over 300 km/h at 900 MHz.

Different from ´he traditional views, Fig. 13 gives a block diagram of a more general and comprehensive concept of wide-sense V2V (WSV2X), which is defined by including not only V2V and V2I communications but also train-to-X (T2X) communications constituted of T2T and train-to-infrastructure (T2I) communications. The idea behind the WSV2X communications is the deployment of wireless communication technology in vehicular and railway networks. In this manner, the vehicles, trains, and infrastructures build up a wireless network enabling them to exchange controlling and traffic information such as road obstacles and accidents via the wireless communication links. Over the past few years, V2X and T2I, respectively, have gained popularity in their attempts to improve road safety and railway safety. However, the development of the T2T channel is not so booming as the other components (V2V, V2I, and T2I) of WSV2X; thus, very limited publications can be found, and many technical challenges still remain in T2T communications. For instance, there is only the T2T channel information at 400 MHz, but not at 900 MHz for GSM-R or 1800 MHz for LTE-R or 2.4 GHz for CBTC.

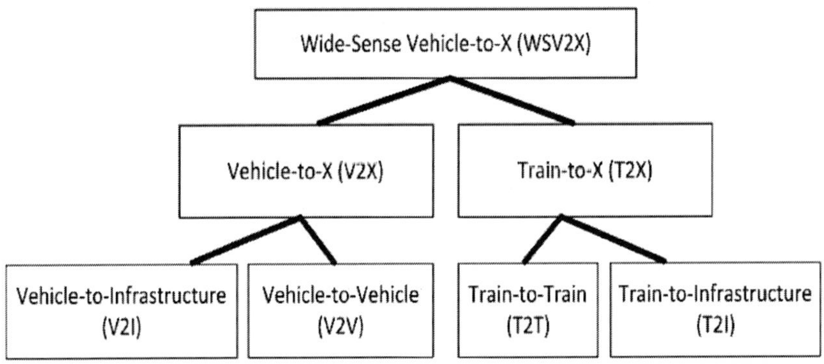

Figure 13. Block diagram of the constitution of the concept of WSV2X.

4.5. Other Key Technologies for Future HSR Development

1) Composite Transportation

The concept of composite transportation has been developed in recent years, which means that the future intelligent transportation network will take the HSR, the subway, and the road as a whole transportation network. For instance, GSM-R network and WiFi technology are used for HSR and subway train-to-ground communications, respectively. However, the next-generation train-to-ground communication system for both HSR and subway will be based on LTE/LTE-A system. Just as the architecture development team from the Research Innovation Technology Administration said in their report [93], wireless communications will be the most enabling technologies in the composite transportation network. How to design the communication network architecture and related key technologies of different layers will be a challenging task.

2) Multipath Birth–Death for Combined Scenarios

Along the HSR lines, there are many combined scenarios, for example, the tunnels and the viaducts, the cuttings and the train stations [94], [95], and the viaducts and the crossing bridges [96]. Serious fading may occur under the combination of these scenarios. Therefore, it is of importance to conduct the research on the multipath birth–death for these combined scenarios. Also worthy of study is the determination of the *demarcation points* of the multipath birth–death for these combined scenarios.

3) SON for Railway

Another advanced and important key technology for LTE-R is the self-organizing network (SON). With this technology, LTE features will be available for managing, configuring, maintaining, and optimizing the LTE-R communication network[97].Intensive research activities are currently running, for example, in the European FP7 project SEMAFOUR[98], for public heterogeneous cellular networks. Within the scope of these projects, the aspects of users with high mobility are investigated. A future research direction for LTE-R will be in the area of developing SON algorithms dealing with specific aspects of HSR. Examples are the optimization of handover parameters, operations of heterogeneous networks consisting of GSM-R and LTE-R, as well as mechanisms to guarantee a robust and reliable operation.

4) Internet of Things for Railway

Sometimes, the debris flow, the broken bridges, the subsidence, and the serious weather conditions will have great influence on the train operation safety. Hence, it is a good idea for future HSR development to equip the railway with Internet of Things to predict and avoid these coming accidents. For instance, Japan has developed the earthquake warning system for her HSR systems [99], [100].

5) Social Network Application for HSR

Social networks such as Facebook, Twitter, LinkedIn, and ResearchGate have been widely used in peoples' everyday life. They facilitate the information transfer and communications between people. We have discussed how to implement social networks for HSR services [101], for example, how to book a train ticket, how to deal with the emergencies on the train, and how to know the tourism attractions and local customs where the train passes. All these can be realized with the utilization of the social network on the train. Then, how to design the architecture and key technologies of such social network and how to realize the intelligent management of the social network on the train remain to be an uneasy task.

6) Vacuum Maglev Train

It is estimated that the moving speeds of the vacuum maglev train is from 600 to 3000 km/h. The pipeline carrying the train is considered to be vacuum in theory, although, we should note that this is often an ideal condition. Then, designing the key technologies for such a new system and understanding the channel conditions in practice remain to be a challenging task [102].

7) Security for HSR

This will definitely be a key point for future development of HSR. The security of the train control data is very crucial, particularly for the WiFi-technology-based train control data transmission system, where the unlicensed frequency band is used. Up to now, there has been few research on the security of train operation control data and the user information security for HSR [103].

8) Fifth-Generation Communications for HSR

The research for the 5G communications system is now on its way. The European Union, the United States, Korea, China, and Japan have developed their organizations for 5G development. HSR scenarios have been recognized as one of the typical scenarios for 5G [60]. Therefore, related key technologies such as massive MIMO technique [104] and the corresponding channel modeling should be paid more attention.

5. CONCLUSION

HSR is developing very quickly, particularly in France, Germany, Japan, Spain, and China. The United States is ready to develop its first HSR in California. Safety, reliability, high efficiency, environment friendliness, comfort, and humanization are the goals of future HSR development. For this application background, we described in this paper some challenging key points on wireless communications for HSR scenarios.

The differences between communications under HSR scenarios and that of public land mobile communication networks were discussed in detail. Particularly for the broadband communication systems, there are still many constraints such as the required working frequency band versus very limited frequency resource, the nonstationary channel and the Doppler effects deteriorating the performance of OFDM-based wireless access methods, the LOS sparse channel restricting the benefit from MIMO techniques, and complex propagation scenarios complicating the mobile channel. Hence, to provide the broadband wireless services in such a high-mobility environment, additional research is required, such as high-resolution analysis in the time/space/frequency domain, quantification of propagation path life and evolving path trajectories, coherence of environments, dynamics of channel responses, handover schemes, and fast power control under high-mobility environments. These are very meaningful for future HSR development.

REFERENCES

1. ETCS/GSM-R Quality of Service—Operational Analysis, ERTMS Std. 04E117, 2010.

2. B. Ning, T. Tang, H. Dong, D. Wen, D. Liu, S. Gao and J. Wang "An introduction to parallel control and management for high-speed railway systems" IEEE Trans. Intell. Transp. Syst., vol. 12, no. 4, pp. 1473-1483, *2011*

3. Z. Li, F. R. Yu, N. Bin and T. Tao "Handoff performance improvements in MIMO-enabled communication-based train control systems" IEEE Trans. Intell. Transp. Syst., vol. 13, no. 2, pp. 582-593, *2012*

4. R. Toledo-Moreo, D. B??taille and F. Peyret "Lane-level integrity provision for navigation and map matching with GNSS, dead reckoning, and enhanced maps" IEEE Trans. Intell. Transp. Syst., vol. 11, no. 1, pp. 100-112, *2010*

5. S. Atev, G. Miller and P. Papanikolopoulos "Clustering of vehicle trajectories" IEEE Trans. Intell. Transp. Syst., vol. 11, no. 3, pp. 647-657, *2010*

6. D. B??taille and R. Toledo-Moreo "Creating enhanced maps for lane level vehicle navigation" IEEE Trans. Intell. Transp. Syst., vol. 11, no. 4, pp. 786-798, *2010*

7. E. Masson, Y. Cocheril, P. Combeau, L. Aveneau, M. Berbineau, R. Vauzelle and E. Fayt"Radio wave propagation in curved rectangular tunnels at 5.8 GHz for metro applications"Proc. 11th ITST, pp. 81-85, *2011*

8. D. T. Fokum and V. S. Frost"A survey on methods for broadband Internet access on trains"EEE Commun. Surveys Tuts., vol. 12, no. 2, pp. 171-185, *2010*

9. IEEE Standard Association. IEEE 802.11: Wireless LANs, IEEE Std.802.11, 2012.

10. S. Karim, S. Khan, A. A. Qureshi and I. Daud"Turning need into demand for Wi-Fi broadband Internet access on trains"Int. J. Comput. Appl., vol. 53, no. 4, pp. 29-42, *2012*

11. LTE communication technologies for automated driving and rail control

12. K. Guan, Z. D. Zhong and B. Ai"Assessment of LTE-R using high speed railway channel model"Proc. ICCMC, pp. 461-464, *2011*

13. Preliminary Document for TG 4P, IEEE Std. P802.15-15-13-0097-04-004p, 2013.

14. T. S. Rappaport, J. N. Murdock, D. G. Michelson and R. Shapiro"An open-source archiving system: RF-propagation measurements and radio-channel modeling and simulation"IEEE Veh. Technol. Mag., vol. 6, no. 2, pp. 24-32, *2011*

15. D. W. Matolak, Q. Wu and I. Sen"5 GHz band vehicle-to-vehicle channels: Models for multiple values of channel bandwidth"IEEE Trans. Veh. Technol., vol. 59, no. 5, pp. 2620-2625, *2010*

16. K. N. Sivertsen, A. Liou and D. G. Michelson"Depth and rate of fading on fixed wireless channels between 200 MHz and 2 GHz in suburban macrocell environments"IEEE Trans. Antennas Propag., vol. 58, no. 10, pp. 3353-3362, *2010*

17. X. Cheng, C. X. Wang, D. I. Laurenson, S. Salous and A. V. Vasilakos"An adaptive geometry-based stochastic model for non-isotropic MIMO mobile-to-mobile channels"IEEE Trans. Wireless Commun., vol. 8, no. 9, pp. 4824-4835, *2009*

18. K. Guan, Z. D. Zhong, B. Ai and T. K??rner"Deterministic propagation modeling for the realistic high-speed railway environment"Proc. IEEE 77th VTC-Spring, pp. 1-5, *2013*

19. A. Richter, M. Landmann and R. Thom??"RIMAX???A maximum likelihood framework for parameter estimation in multidimensional channel sounding"Proc. ISAP, pp. 53-56, *2004*

20. WINNER II channel models*2008*

21. B. Ai, Z. X. Yang, C. Y. Pan, J. H. Ge, Y. Wang and Z. Lu"On the synchronization techniques for wireless OFDM systems"IEEE Trans. Broadcast., vol. 52, no. 2, pp. 236-244, *2006*

22. B. Ai, J. H. Ge and Y. Wang"Symbol synchronization technique in COFDM systems"IEEE Trans. Broadcast., vol. 50, no. 1, pp. 56-62, *2004*

23. B. Ai, Y. Shen, Z. D. Zhong and B. H. Zhang"Enhanced sampling clock offset correction"IEEE Trans. Consum. Electron., vol. 57, no. 2, pp. 696-704, *2011*

24. M. W. Wen, X. Cheng, B. Ai, X. Wei and A. P. Huang"High precision multi-cell channel estimation algorithm based on successive interference cancellation"Proc. 3rd CMC, pp. 385-388, *2011*

25. J. Yao, S. S. Kanhere and M. Hassan"Mobile broadband performance measured from high-speed regional trains"Proc. IEEE VTC Fall, pp. 1-5, *2011*

26. D. W. Matolak and O. Athens"Channel modeling for vehicle-to-vehicle communications"IEEE Commun. Mag., vol. 46, no. 5, pp. 76-83, *2008*

27. R. He, Z. Zhong, B. Ai, L. Xiong and H. Wei"A novel path loss model for high-speed railway viaduct scenarios"Proc. 7th WiCOM, pp. 1-4, *2011*

28. H. Wei, Z. Zhong, L. Xiong, B. Ai and R. He"Study on the shadow fading characteristic in viaduct scenario of the high-speed railway"Proc. 6th ICST Commun. Netw. China, pp. 1-5, *2011*

29. M. Hata"Empirical formula for propagation loss in land mobile radio services"IEEE Trans. Veh. Technol., vol. VT-29, no. 3, pp. 317-325, *1980*

30. R. S. He, Z. D. Zhong, B. Ai, J. Ding and Y. Yang"Propagation measurements and analysis of fading behavior for high speed rail cutting scenarios"Proc. IEEE GLOBECOM, pp. 5015-5020, *2012*

31. K. Guan, Z. D. Zhong, B. Ai and T. K??rner"Measurements and modeling of crossing bridges on high-speed railway at 930 MHz"IEEE Trans. Veh. Technol., vol. 63, no. 2, pp. 502-517, *2014*

32. G. Gougeon, Y. Lostanlen and L. Maviel"Coupling a deterministic propagation model with diffuse scattering and urban furniture for small cells" Proc.5th EuCAP, pp. 3448-3452, *2011*

33. LTE-Evolved Universal Terrestrial Radio Access (E-UTRA)—Physical Layer Procedures, 3GPP Std. TS 36.213, V9.2.0 Release 9, 2010.

34. Requirements for theRadio Interface(s) for International MobileTelecommunications- 2000 (IMT-2000), ITU Std. Rec. ITU-R M: 1034-1, 2000.

35. Universal Mobile Telecommunications Systems (UMTS): Selection Procedures for the Choice of Radio Transmission Technologies of the UMTS, ETSI Std. TR 101 112 v3.2.0, 1998. *1998*

36. B. Ai, R. S. He, Z. D. Zhong, K. Guan, B. H. Chen, P. Y. Liu and Y. X. Li"Radio wave propagation scene partitioning for high-speed rails"Int. J. Antennas Propag., vol. 2012, pp. 815232-1-815232-7, *2012*

37. L. Liu, C. Tao, J. H. Qiu, H. J. Chen, L. Yu, W. H. Dong and Y. Yuan"Position-based modeling for wireless channel on high-speed railway under a viaduct at 2.35 GHz"IEEE J. Sel. Areas Commun., vol. 30, no. 4, pp. 834-845, *2012*

38. R. S. He, Z. D. Zhong, B. Ai, G. Wang, J. Ding and A. F. Molisch"Measurements and analysis of propagation channels in high-speed railway viaducts"IEEE Trans. Wireless Commun., vol. 12, no. 2, pp. 794-805, *2013*

39. B. Ai"Interview"Electron. Lett., vol. 47, no. 21, p. 1158, *2011*

40. R. S. He, Z. D. Zhong, B. Ai and J. Ding"An empirical path loss model and fading analysis for high-speed railway viaduct scenarios"IEEE Antennas Wireless Propag. Lett., vol. 10, pp. 808-812, *2011*

41. R. S. He, Z. D. Zhong, B. Ai and J. Ding"Measurements and analysis of the directional antenna bottom area in high speed rail"Proc. IEEE APSURSI, pp. 1-2, *2012*

42. R. S. He, A. F. Molisch, Z. D. Zhong, B. Ai, J. Ding, R. Chen and Z. Li"Measurement based channel modeling with directional antennas for high-speed railways"Proc. IEEE WCNC, pp. 2932-3936, *2013*

43. R. S. He, Z. D. Zhong, B. Ai, K. Guan and R. Chen"Simplified analytical propagation model for railway environments based on uniform theory of diffraction"Electron. Lett., vol. 49, no. 6, pp. 397-399, *2013*

44. R. S. He, Z. D. Zhong, B. Ai and J. Ding"Measurements and analysis of short-term fading behavior for high-speed rail viaduct scenario"Proc. IEEE ICC, pp. 4563-4567, *2012*

45. R. S. He, Z. D. Zhong, B. Ai, J. Ding, Y. Yang and A. F. Molisch"Short-term fading behavior in high-speed railway cutting scenario: Measurements, analysis, and statistical models"IEEE Trans. Antennas Propag., vol. 61, no. 4, pp. 2209-2222, *2013*

46. R. S. He, Z. D. Zhong, B. Ai and J. Ding"Propagation measurements and analysis for high-speed railway cutting scenario"Electron. Lett., vol. 47, no. 21, pp. 1167-1168, *2011*

47. F. Abrishamkar and J. Irvine"Comparison of current solutions for the provision of voice services to passengers on high speed trains"Proc. IEEE 52nd VTC Fall, pp. 1498-1505, *2000*

48. R. S. He, Z. D. Zhong and C. Briso"Broadband channel long delay cluster measurements and analysis at 2.4 GHz in subway tunnels"Proc. IEEE 73rd VTC Spring, pp. 1-5, *2011*

49. R. He, Z. D. Zhong, B. Ai, K. Guan, B. Chen, J. I. Aionso and C. Briso-Rodriguez"Propagation channel measurements and analysis at 2.4 GHz in subway tunnels"IET Microw., Antennas Propag., vol. 7, no. 11, pp. 934-941, *2013*

50. K. Guan, Z. D. Zhong, B. Ai, C. Briso-Rodriguez and J. I. Alonso"Measurement of distributed antenna systems at 2.4 GHz in a realistic subway tunnel environment"IEEE Trans. Veh. Technol., vol. 61, no. 2, pp. 834-837, *2012*

51. K. Guan, Z. D. Zhong, B. Ai, R. He and C. Briso-Rodriguez"Five-zone propagation model for large-size vehicles inside tunnels"Progr. Electromagn. Res., vol. 138, pp. 389-405, *2013*

52. K. Guan, Z. D. Zhong, B. Ai and C. Briso-Rodriguez"Measurement and modeling of subway near shadowing phenomenon"Proc. CHINACOM, pp. 1-5, *2010*

53. Y. P. Zhang and Y. Hwang"Enhancement of rectangular tunnel waveguide model"Proc. APMC, pp. 197-200, *1997*

54. K. Guan, Z. D. Zhong, B. Ai and C. Briso-Rodriguez"Novel hybrid propagation model inside tunnels"Proc. 75th IEEE VTC Spring, pp. 1-5, *2012*

55. K. Guan, Z. D. Zhong, B. Ai and C. Briso-Rodriguez"Propagation mechanism analysis before the break point inside tunnelsProc. 74th IEEE VTC Fall, pp. 1-5, *2011*

56. K. Guan, Z. D. Zhong, B. Ai and C. Briso-Rodriguez"Propagation mechanism modeling in the near region of circular tunnels"IET Microw., Antennas Propag., vol. 6, no. 3, pp. 355-360, *2012*

57. 57. K. Guan, Z. D. Zhong, B. Ai and C. Briso-Rodriguez"Modeling of the division point of different propagation mechanisms in the near-region within arched tunnels"Wireless Pers. Commun., vol. 68, no. 3, pp. 489-505, *2011*

58. 58. K. Guan, Z. D. Zhong, B. Ai, R. He, Y. Li and C. Briso-Rodriguez

59. "Propagation mechanism modeling in the near-region of arbitrary cross-sectional tunnels"Int. J. Antennas Propag., vol. 2012, pp. 183145-1-183145-11, *2012*